J.M. fgure

Advances in Anatomy
Embryology and Cell Biology

Vol. 162

GW00697227

Springer

Berlin
Heidelberg
New York
Barcelona
Hong Kong
London
Milan
Paris
Tokyo

K. Punkt

Fibre Types
in Skeletal Muscles

With 43 Figures and 6 Tables

Springer

KARLA PUNKT, PD Dr. rer. nat. habil
University of Leipzig
Faculty of Medicine
Institute of Anatomy
Liebigstr. 13
04103 Leipzig, Germany

e-mail: punktk@medizin.uni-leipzig.de

ISSN 0301-5556
ISBN 3-540-42603-5 Springer-Verlag Berlin Heidelberg New York

Library of Congress-Cataloging-in-Publication-Data
Punkt, K. (Karla), 1949– . Fibre types in skeletal muscles / K. Punkt. p. cm.
– (Advances in anatomy, embryology and cell biology ; vol. 162)
Includes bibliographical references and index.
    ISBN 3540426035 (pbk.)
1. Muscles. 2. Muscle cells. I. Titel. II. Series.
QP321 .P86 2001   611'.0186–dc21

Springer-Verlag a member of BertelsmannSpringer
Science + Business Media GmbH

http://www.springer.de

© Springer-Verlag Berlin Heidelberg 2002
Printed in Germany

Production: PRO EDIT GmbH, 69126 Heidelberg, Germany
Printed on acid-free paper – SPIN: 10847501   27/3130wg - 5 4 3 2 1 0

Dedicated to my children,
Anita and Torsten

Preface

Worldwide, numerous textbooks and publications have dealt with research on muscle fibres carried out under different points of view. In addition, comprehensive works such as *Myology* (Engel and Franzini-Armstrong 1994), *Disorders of Voluntary Muscle* (Walton et al. 1994), and *Skeletal Muscle* (Schmalbruch 1985) as a volume of the work *Handbook of Microscopic Anatomy*, have been published. Moreover, proceedings from myology symposiums give us access to the present state of the art in muscle research. The book *The Dynamic State of Muscle Fibres* (Pette 1990a) summarizes the contributions to the symposium of the same name, which was held in Constance in 1989.

Considering these outstanding works one has to ask the question: Why do we need the present book?

The first reason is that results from ongoing research expand scientific knowledge continuously. When dealing with muscle research one soon realizes that muscle tissue is a fascinating subject, whose secrets have not yet been revealed completely. The application of new techniques in muscle fibre research enables and provokes us to go deeper into the nature of muscle tissue. The results are findings that add a new dimension to what is already known. For instance, the detailed metabolic characterization of muscle fibre types in the context of an intact histological section has been performed only recently using cytophotometrical quantification of enzyme activities.

The second reason for this book is of a more pragmatic nature. In my opinion, researchers, physicians, and students who are interested but not specialized in myology will be pleased to have a concise summary of our present knowledge on muscle fibres. After all, 109 pages are read more quickly than thousands!

The present book aims at characterizing fibre types of skeletal muscles from different current points of view. For this purpose results from our own research have been compared to and combined with the results found in the literature. A main emphasis of this review is put on fibre typing by quantitative enzyme histochemistry as well as the adaptability of defined fibre types to altered physiological and pathological conditions. We show that the fibre metabolism is very flexible under changing conditions. This impressively demonstrates the plasticity of muscle fibres.

I am very grateful to Professor Schiebler who encouraged me to write this review of the current knowledge on skeletal muscle fibres. If this book can offer the reader a general overview about the properties of the different fibre types, and if it perhaps inspires the reader to further work on muscles, then it will have served its purpose.

Leipzig, November 2001   K. PUNKT

Contents

Acknowledgements

I wish to thank Professor Katharina Spanel-Borowski, head of the Institute of Anatomy of the University of Leipzig, for providing good working conditions which are necessary to produce a book like this.

Abbreviations

AB	Antibody
ATP	Adenosine triphosphate
ATPase	Adenosine triphosphatese
EDL muscle	Extensor digitorum longus muscle
EGb	Extract Ginkgo biloba
EOM	Extraocular muscle
FF	Fast-twitch-fatiguable
FG	Fast-glycolytic
FOG	Fast-oxidative glycolytic
FR	Fast-twitch-fatigue-resistent
GAST muscle	Gastrocnemius muscle
GPDH	Glycerol-3-phosphate dehydrogenase
HMM	Heavy meromyosin
IgG	Immunoglobulin G
LDH	Lactate dehydrogenase
LMM	Light meromyosin
MDF	Myogenic determination factor
MHC	Myosin heavy chain
NBT	Nitro blue tetrazolium
NOS	Nitric oxide synthase
PKC	Protein kinase-C
RNA, mRNA	Ribonucleic acic, messenger RNA
S	Slow-twitch
SDH	Succinate dehydrogenase
SO	Slow-oxidative
SOL muscle	Soleus muscle
SR	Sarcoplasmic reticulum
STZ	Streptozotocin
TnT	Troponin T
VAST muscle	Vastus lateralis muscle

1 Introduction

1.1
General Remarks

Skeletal muscle tissue is an extremely heterogeneous tissue consisting of a variety of fibre types. A prominent characteristic of skeletal muscles is the ability to adapt to changed functional conditions by varying their fibre properties. The investigation of these fibre properties has been at the centre of interest for about 125 years. According to the techniques available at the relevant time, the differentiation of muscle fibres has been developed from differences in their colour, red and white (Ranvier 1873) up to differences in their gene expression (Pette 1990a). Our own studies deal with the metabolic characteristic of fibre types as reported in Chaps. 4 and 5.

The definitive function cells of skeletal muscles are the intra- and extrafusal fibres. The intrafusal fibres have a sensoric function; the extrafusal fibres are responsible for the motoricity. The extrafusal fibres, in the following called muscle fibres, are the topic of the present book. Muscle fibres form a heterogeneous fibre population and differentiate in their physiological, biochemical and morphological properties. This is due to the various forces and movements generated and developed by the skeletal muscle. Muscle fibres are innervated by α-motoneurons. Their pericarya lie in the anterior horn of the spinal cord and in the motoric cranial nerve nuclei respectively. Their axons run to the skeletal muscles, where they innervate distinct muscle fibres by motoric endplates. The number of muscle fibres innervated from one α-motoneuron varies greatly. It depends on the movements which are demanded of the relevant muscle. For example, one α-motoneuron innervates only three to five fibres in extraocular muscles, whereas about 1,000 fibres are innervated by one α-motoneuron in limb muscles. The α-motoneuron and all of its innervated muscle fibres form a motor unit, which is the basis of neuromuscular function. The smaller the motor units of a muscle, the more precise movements are possible. The motoneuron determines the properties of the muscle fibres which are innervated by it. The muscle fibres belonging to the same motor unit have similar morphological and physiological properties. However, the metabolic properties of fibres comprising a motor unit can vary (Larsson 1992, this study). Moreover, it was found in several rat muscles that within one motor unit different isoforms of myosin heavy chains are expressed (Schiaffiano et al. 1990; Larsson et al. 1991). This is interpreted as a reflection of transformation processes or as motoneuron dysfunction. Jostarndt et al. (1996) identified different subtypes of slow twitch muscle fibres in human skeletal muscles which differ in their expression of myosin light chain isoforms. According to their physiological characteristics, three types of motor units are distinguished: one type with slow twitch muscle fibres and two types with fast twitch muscle fibres. This means that the physiological classification of muscle fibres into slow and fast corresponds to the classification of the motor unit (Edström and Kugelberg 1968). The physiological properties of the comprised muscle fibres designate the motor units as FF (fast-twitch fatigable), FR (fast-twitch fatigue-resistant) and S (slow-twitch), Burke (1983).

1.2
Structure and Contraction Mechanism of Muscle Fibres

Skeletal muscle fibres are multinucleated; they are syncytia of fused mononucleated myoblasts. Vertebrate skeletal muscle fibres are 10–100 μm in diameter and several centimeters long. The fibres of human muscles can be up to 20 cm long. The fibre length in the soleus muscle of the adult rat were measured to be 12 mm, and in the extensor digitorum longus muscle, it is 10 mm (Close 1964). The nuclei of muscle fibres are located under the sarcolemma (plasmalemma of the muscle cell); they were pushed aside after the development of myofibrils. The muscle fibres are packed with numerous myofibrils – they occupy more than 80% of the volume of an adult muscle fibre. Each myofibril is enveloped by sarcoplasmatic reticulum (SR) and runs parallel to the fibre axis. Myofibrils consist of thin and thick contractile filaments which contain the proteins actin and myosin, respectively. The contractile filaments are overlappingly arranged within the myofibril, producing a regular pattern of transverse stripes. The fibrils are in register, thus producing a striated appearance in the fibre as a whole. The cross striation is a pattern of bands consisting of thick or thin filaments or both. The ultrastructure and the molecules of a cross striated muscle fibre is schematically demonstrated in Fig. 1. The repeating unit of the banding pattern is the sarcomere which extends from Z line to Z line. The sarcomere is the fundamental contractile unit of striated muscle. The knowledge about the structure of the sarcomere was the condition for understanding the molecular mechanism of muscular contraction. Each Z line bisects a lighter I band (isotropic in polarized light). The I band is formed by thin actin filaments which are bound by connecting proteins (A-actinin, Z-protein, vinculin) to the Z line. The actin filaments run from their attachment sites at the Z line through the I band into the A band (anisotrop in the polarized light), where they overlap with the thick myosin filaments. In the middle of each A band is a lighter H zone, owing to the absence of overlapping thin filaments. Only thick filaments are present in the H zone. It contains a darker region called the M zone. Here, the thick filaments are interconnected by cross bridges. Cross sections through the A band show hexagonal patterns: In the overlap zone of A band the thick filaments form a hexagonal array, and each thick filament is surrounded by six thin filaments. In the H zone the hexagonal array of the thick filaments is shown.

Thick filaments of vertebrate skeletal muscle consist exclusively of myosin. The length of the molecule is 140–170 nm. It consists of a rod-shaped part, 2 nm thick and 150 nm long, and a globular part, 5 nm thick and 10–20 nm long (Schmalbruch 1985c), which is composed of two subunits. The myosin molecule is a hexamer (molecular weight about 470,000 Da), consisting of two similar heavy chains (HMM, molecular weight 200,000 Da) and four light chains (LMM, each of molecular weight about 20,000 Da). The two HMM wound around each other forming a two-chain

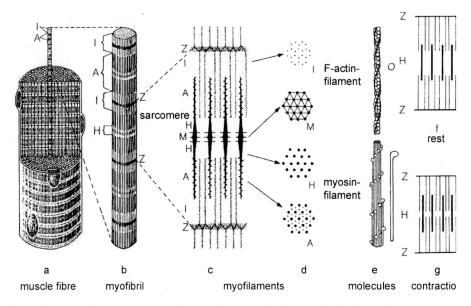

Fig. 1. Structure of a skeletal muscle fibre. **a** Single muscle fibre. **b** Myofibril according to electron micrographs. The zones of cross striation are marked by Z, I, A, H, and M. **c** Schematic ultrastructure of one sarcomere. Thin actin filaments and thick myosin filaments are arranged in register. **d** Cross-sectional appearance of sarcomere at different segments (I, M, H, A). **e** Molecular structure of F-actin and myosinfilaments. **f** Sarcomere in the resting state. **g** Sarcomere in the contracting state. (According to Schiebler 1999)

α-helix. The helix corresponds to the rod-shaped part of the molecule. In the amino terminal each of the two HMM folds to form one globular head. The globules are the S1 subfragments, whereas the rod-shaped tail of HMM, the "hinge," is the S2 subfragment. Each S1 subfragment contains two LMM which are of two chemically distinct classes. One of each class is associated with each head. HMM of the head carries the ATPase activity of myosin. The heads interact to actin, forming cross bridges. LMM is the dominant part of the rod-shaped tail of the molecule. LMM is of simple rod-shaped structure and retains the ability of the myosin molecules to assemble themselves at suitable strength into filaments. The thick filament is a polymer of myosin, containing about 300 molecules. The molecules initially associate tail-to-tail, in an antiparallel array, and the filaments elongate by staggered head-to-tail parallel additions of molecules at each end. The myosin tails form the core of the filament, while the heads lay at the surface. The thick filaments of vertebrates contain, in addition to myosin, small amounts of additional proteins, whose functions are not completely understood (for details, see Craig 1994). Thick filaments of vertebrates are 15 nm in diameter, and their length is 1.6 µm corresponding to the length of A band (Schmalbruch 1985c).

Thin filaments of vertebrate skeletal muscle consist predominantly (70%) of actin. Actin monomers are globular (G actin); the molecular weight is 41,800 Da. In muscles, actin exists as a polymer (F actin), containing approximately 360 molecules in vertebrates. G actin monomers form a double-stranded helix within the F actin filament. The regulatory protein troponin is attached at regular intervals to the helix. The other regulatory protein tropomyosin lies in the grooves of the helix. A fourth component of actin is the protein nebulin which interacts with actin and tropomyosin. Thin filaments of vertebrates are about 1 μm long and 68 nm in diameter (Schmalbruch 1985c).

The muscle fibre is not just a tube filled with myofibrils and aqueous cytoplasm. As mentioned above, intermediate filaments exist at the Z line. A sarcomere-related system links the sarcolemma and also the interstitial connective tissue to the myofibrils. The intermediate filament systems within the myofibrils contain mainly titin filaments. For details about the cytoskeletal elements and adhesion molecules see Franzini-Armstrong and Fischman (1994) and Knudsen and Horwitz (1994).

The knowledge of the ultrastructure of the muscle fibre had led to the theory of sliding filaments to explain the contractile mechanism. H. E. Huxley and Hanson (1954) and A.F. Huxley and Niedergerke (1954) independently found out that contraction was a result of the sliding of filaments past each other, producing greater overlaps of the filaments without a change the lengths of the filaments themselves. The length of the A band remains constant during contraction, whereas the I band and H zone are shortened (Fig. 1f,g). The filaments remain constant in length when the muscle contracts. Contraction involves the stepwise interaction of the myosin heads with the actin monomers of the thin filaments. Links between thick filaments and thin filaments are formed in the region of overlap. These cross bridges provide the necessary mechanical link between thick and thin filaments, and the sliding force is generated by a change in conformation of the bridges while attached to actin. The thin filaments thereby are pulled into the thick filament array. Cross bridges act cyclically: attaching to actin, changing conformatin, detaching, then attaching again further along the filament and repeating the cycle. The interacting filaments develop force, and they slide along each other. The distance between Z lines decreases, and the sarcomere shortens. The interaction of myofilaments is controlled by Ca^{2+} ions stored in and released from the sarcoplasmic reticulum. The electrical signal for Ca^{2+} release to initiate contraction is conducted by a transverse tubular system (T system) from the sarcolemma into the fibre. The T tubules are invaginations of the sarcolemma; they make contact with the SR at the triadic junctions. Here, the depolarization of the sarcolemma is conducted to the SR, followed by the release of calcium ions from the cisterns of SR. The T tubules are located along the myofibrils at the A-I junctions and make the concerted contraction of the whole muscle fibre possible. Free calcium ions bind to troponin, inhibiting the depressor action of both regulatory proteins troponin and tropomyosin. In this way, the muscle contraction is initiated. Myofibrils are mechanically linked across the sarcolemma over reticular fibrils to the collagen fibrils of tendons. By this, the contraction force is transmitted. Relaxation involves the re-uptake of calcium into the cisterns of SR and disengagement of the myosin heads from their binding sites at the actin monomers. The energy for contraction and also the energy necessary to store Ca^{2+} ions within the sarcoplasmic reticulum against a concentration gradient is obtained from splitting ATP by ATPase. ATP is produced by oxidative phosphorylation within mitochondria, and also by glycolysis within the cytosol. Both the myosin heads and the reticulum membranes are sites of ATPase activity, called myosin ATPase and sarcoplasmic ATPase, respectively.

According to the contracting velocity of its myofibrils a muscle fibre is determined as slow or fast twitch fibre. The skeletal muscle fibres of mammals are almost twitch fibres; they conduct action potentials and give an all-or-nothing response. They are innervated at only one endplate. In extraocular muscles and other muscles of the head region of mammals and in muscle spindles also other slowly contracting fibres occur. These are slow tonic fibres which are innervated at multiple sites along their length. Slow tonic fibres are common in avian and amphibian muscles. These specific fibre types are described in Chap. 12.

The contractile properties differ in slow and fast fibre types. This is expressed in their morphology, e.g. by the abundance or scarcity of membranes of SR and T system, and of mitochondria. There are also differences in the structure of myofibrils, the M and Z lines, the motor endplates and in myosin isoforms. Mainly the different myosin isoforms determine the fibre diversity in skeletal muscles. Chapters 3 and 4 deal with the theoretical background of fibre diversity.

1.3
Intrafusal Fibres of Muscle Spindles

Apart from ordinary (extrafusal) muscle fibres responsible for muscle contraction, there are intrafusal muscle fibres which have a sensory function. Intrafusal fibres lie as bundles within a spindle-shaped capsule. Such muscle spindles are mechanoreceptors sensitive to muscle length and changes in it. Intrafusal fibre bundles are arranged in parallel with extrafusal fibres, their ends attached to connective tissue, tendon or extrafusal endomysium. They receive both a motor and a sensory innervation. The sensory innervation, which responds to active and passive changes in muscle length, occupies the equatorial region of intrafusal bundles. The motor innervation is distributed to the polar regions that extend on each side. Activation of the motor innervation elicits contractions in the polar regions that modify the sensory discharge. The intrafusal fibres elicit responses from sensory nerve endings by active contraction or by passively transmitting tension from the connective tissue of the muscle. In other words, the equatorial regions (sensory region), where the sensory endings terminate are non-contractile, while the polar regions (motor region) contract actively.

In average, one muscle spindle contains four to ten intrafusal fibres. The structure of the muscle spindle and the number of containing intrafusal fibres differ in different species. This chapter deals with mammalian spindles.

Two types of intrafusal muscle fibres can be recognized on the basis of differences in length, diameter and equatorial nucleation: nuclear bag fibres and nuclear chain fibres. Nuclear bag fibres are thick fibres which show an equatorial accumulation of myonuclei. Nuclear chain fibres are thin fibres with a central chain of myonuclei. Histochemical studies revealed that there are two kinds of bag fibres (see review by Barker 1974), designated as bag_1 and bag_2 by Ovalle and Smith (1972). This means that three types of intrafusal fibres are distinguished using histochemistry: nuclear chain fibres, bag_1 and bag_2 fibres. The three fibre types differ in their glycogen content and in their activities of myofibrillic ATPase, phosphorylase and nicotinamide adenine dinucleotide tetrazolium reductase (for references, see Barker and Banks 1994). The differences in the histochemical staining pattern between intrafusal fibre types are less clear than those between extrafusal fibre types, because the histochemical profiles of intrafusal muscle fibres are subject to regional variations. There are regional distribution differences for the different myosin isoforms and other molecules (for review see Soukup et al. 1995) within the same intrafusal muscle fibre type. This indicates a more complex molecular situation in the intrafusal fibres than in the extrafusal ones. Soukup et al. (2000) reported that the three types of intrafusal fibres are unique in coexpressing several MHC isoforms, including special spindle-specific ones, such as slow tonic (slow developmental) and α cardiac-like MHC. Embryonic and neonatal MHC which are typical for muscle development have been also found in intrafusal fibres of adults (for review, see Soukup et al. 1995; Walro and Kucera 1999). It was reported (Soukup et al. 1995) that in the rat, each intrafusal fibre type has a typical MHC pattern. It comprises at least 6 MHC isoforms in nuclear bag$_2$ fibres (embryonic, neonatal, slow twitch/β cardiac, α cardiac-like, slow tonic/slow developmental and fast twitch), 4 MHC isoforms in nuclear bag$_1$ fibres (embryonic, slow twitch/β, slow tonic/slow developmental and α cardiac-like) and 2 MHC isoforms in nuclear chain fibres (neonatal and fast twitch).

Intrafusal muscle fibre types distinguish from each other by their distinct MHC pattern and expression of spindle-specific MHC isoforms. These MHC isoforms are not expressed in extrafusal fibres; therefore intrafusal fibres distinguish also from extrafusal fibres in all mammalian and human muscles (for review see Soukup et al. 1995, Walro et al. 1999). The unique expression of slow tonic, α cardiac-like, embryonic and neonatal MHCs in limb muscles of adult mammals is restricted to intrafusal fibres.

1.4
Satellite Cells

Satellite cells are mononucleated cells lacking myofibrils and situated beneath the basal lamina of skeletal muscle fibres. They arise from myogenetic cells during embryomyogenesis, but they are not equivalent to myogenic cells in early embryos (Bischoff 1994). Satellite cells function as myogenic stem cells in mature muscles and have an impressive capacity for growth and self-maintenance. The satellite cell shows pronounced mitotic activity in growing, denervated or injured muscles. Two closely associated heterochromated nuclei, of which one is a myonucleus and one a satellite cell nucleus, are often found in growing and diseased muscles (Moss and Leblond 1971). These two nuclei are probably the result of satellite cell mitosis; one of the daughter cells fuses with an intact part of the myofibre, and one remains a proliferating satellite cell. In this way, skeletal muscle fibres can regenerate after disease. Myonuclei lost as a result of injury can be replaced only by proliferation and fusion of satellite cells. For more details about satellite cells see Bischoff (1994).

2 When Does Differentiation of Muscle Fibres Begin and Which Factors Characterize a Fibre Type?
How Can the Muscle Fibre "Know" Whether to Be Fast or Slow, and Since When?

The origin of muscle fibre diversity already lies in the earliest phase of skeletal muscle development. On the level of transcription, myogenic determination factors (MDFs) are able to convert mesodermal cells into skeletal myoblasts. Myoblasts express at least one of the MDFs and occur in a number of types, which are changing during development. The activation of gene coding for muscle-specific proteins happens nearly at the same time as the fusion of myoblasts to myotubes. The synthesis of the several contractile and other proteins is coordinated and controlled on the transcription level. Muscle proteins exist in a large variety of isoforms which change during development. Many of the isoforms, e.g. myosin heavy chain (MHC) isoform, are products of multigene families. Others, such as troponin T (TnT), derive from differential RNA splicing. On the gene level, activation is sequential. That means, the activation of the transcription of one gene is paralleled by the suppression of another gene which was previously active. That mechanism is not fully understood, because the specific properties and the subtle differences of most of the isoforms are unknown. However, in the case of myosin, properties are well-known. The maximum speed of shortening indicates the function of myosin. A correlation has been found between contraction properties of myosin and the MHC composition. Moreover, the MHCs can be taken as suitable markers for the investigation of the diversity of muscle fibres.

In earlier studies only two MHCs – slow and fast – were found; today we know at least 16 MHC isoforms. In mammals, each MHC isoform is a product of a distinct gene. Thus, the "knowledge" of a muscle fibre on whether it has to be slow or fast derives from the genetic code. It is worth noting that the innervation of the fibre does not play a role in the development of fibre types. The fibre diversity is an intrinsic property of skeletal muscle. This was acquired from studies involving both chicks and rats in which denervation was effected in the embryo before the nerve penetreated limb muscle primordia (Butler et al. 1982; Phillips and Bennet 1984; Fredette and Landmesser 1991; Condon et al. 1990). These studies showed that the absence of the nerve does not inhibit the emergence of fibre diversity, although the fetal muscle growth is dramatically effected.

At the molecular level the differentiation to fibre types involves the programmed switching of protein isoforms. In this way, a plan of fibre diversity is introduced early in the histogenesis.

During the maturation of animals, this fundamental plan may be modulated by environmental effects, such as innervation, activity, or hormones. The answer of fibres to these environmental stimuli depends on their initial specification. For example, cross-innervation and chronic stimulation can be followed by fibre type transformation, which is rooted in the expression of a new MHC isoform in the fibre. However, not all fibres converted, and the capacity of fibres to adjust to incoming stimuli was limited. These limitations were described as the "adaptive range" of fibres, suggesting that the fibres do not forget their inaugural program.

At which period of muscle histogenesis does the emergence of diversity appear? For this, the MHC isoforms will be used as molecular markers of differentiation. It was shown that the sequential transitions between MHC isoforms (embryonic, slow, neonatal and adult fast) characterize the dynamic state of muscle development (Kelly and Rubinstein 1994).

Muscle histogenesis is initiated by the differentiation of myotubes. Primary myotubes were generated by fusion of myoblasts (in developing human muscle between 8 and 10 weeks of gestation). Later, a new generation of myoblasts fused with primary myotubes to secondary myotubes. According to sequential ordering of fibre diversity, primary and secondary myotubes synthesize different MHC isoforms. It should be noted that these isoforms do not correlate with those of slow or fast fibres. The relationships between primary and secondary myotubes and slow and fast muscle fibres are extremly complex. Early in the histogenesis the organization of primary myotubes varies from muscle to muscle, e.g. shown for rat soleus (SOL) and extensor digitorum longus (EDL) muscles (Kelly and Rubinstein 1994). The identity of the muscle is already marked in the primordium. The developmental behaviour of both primary and secondary myotubes varies in different muscles. While primary myotubes of SOL express embryonic and slow MHC, primary myotubes of EDL form a heterogeneous population. Some express embryonic and slow MHCs; others express embryonic and perinatal MHCs. In the neonate, all primary fibres of SOL continue to express slow MHC. They become slow fibres in the adult. Many secondary cells which initially accumulate adult fast MHCs have now shifted to slow MHC accumulation. The SOL becomes increasingly specialized as a slow muscle. In contrast, in the EDL many primary fibres and all secondary fibres produced perinatal MHC. They will mature to fast fibres. The decision to produce perinatal MHC and not slow MHC is at the same time a decision for fast MHC. This choice is made before the adult MHCs expressed! That means that early in histogenesis the primary myotubes "know" that they belong to the developing EDL, which is increasingly specialized as a fast muscle. This phenomenon in the primordium of muscles is not yet fully understood.. New questions about the developmental program are continuously arising. Is the muscle diversity due to lineage or nonlineage specification? Or, in other words, is the population of myoblasts heterogeneous or homogenous? Many studies deal with this questions. Probably, the truth lies between these opposing theories.

The step-by-step development of motor units, which involves the development of innervation of myotubes from polyneural to mononeural, causes the mosaic pattern of fibre types.

3   The Difficulties and Possibilities of Classifying Muscle Fibres in Distinct Nonoverlapping Types

There are three properties which differ from fibre type to fibre type: MHC isoforms, TnT isoforms, and metabolic enzymes. As already mentioned in the previous chapter, the MHC isoforms are suitable markers of fibre types. MHCs are discretely distributed among normal adult muscle fibres. Mostly, only one or predominantly one isoform is expressed per fibre in normal adult muscles. In contrast to MHC isoforms, the TnT isoforms are distributed less discretely. Among the fast fibres, "myriad" TnT isoforms exist, often with multiple isoforms in each fibre (Kelly and Rubinstein 1994). Furthermore, there is no clear correlation of TnT isoforms and MHC isoforms. TnT isoforms are not suitable as markers for practical fibre typing. The levels of metabolic enzymes characterize different fibre types. However, these classifications do not correlate with the MHC classification; considerable overlapping exists. Here, it should be emphasized: Muscle fibres can be typed according to several classification systems. The different criteria which underlie different classification systems are not obviously interchangeable with another. The intermingling of different systems have led to some confusions in the literature about muscle fibre types. In the following, an overview about the usual muscle fibre types is given, considering both literature and our own results. Special fibre types which occur, e.g. in extraocular muscles and avian skeletal muscles, are described in Chap. 12.

The characteristics of fibre types are summarized in Table 1, several histochemical pictures are demonstrated for rat EDL in Fig. 1. According to the concerning classification system, muscle fibres can be classified as slow and fast, or as I, IIA, IIB (IIX) or as SO, FOG I, FOG II and FG. The diversity of muscle fibres can be proven by physiological, histochemical, biochemical or cytophotometrical methods. The physiological classification into slow and fast muscle fibres can be demonstrated on the histological section. Using antibodies against slow or fast MHCs, slow or fast fibres are stained (Fig. 2a). Routine histochemical fibre typing is usually basing on the acivity of myofibrillic adenosine triphosphatase (ATPase, Fig. 2b,c) resulting in three fibre types: I, IIA, and IIB. The designation I, IIA, or IIB is the designation of the MHC isoform expressed from the respective fibre. The theoretical background for fibre typing with myofibrillic ATPase (ATPase fibre typing) is as follows: The energy for muscle contraction is derived from splitting of the adenosine triphosphate (ATP). The myosin molecule has ATPase activity whose level differs between the MHC isoforms. Slow MHCs split ATP slowly; fast MHCs split ATP quickly correlating with the maximal shortening velocity of the fibres. According to the different acid and alkali labilities of their ATPase activity, three MHC isoforms can be distinguished: MHC I,

Table 1. Classification of skeletal muscle fibres

Classification system	Fibre types			
Physiology	Slow	Fast		
Different MHCs	MHC I/b Slow splitting of ATP (300 mol/s)	MHCs IIA, IIB, IIX Fast splitting of ATP (600 mol/s)		
Contraction time	90–140 ms	40–90 ms		
Fatigue resistance	High (Endurance)	Low (Fast force)		
Maximal tetanic force/ motor unit	350 mN	550–1300 mN		
Frequency of stimulation	5–25 Hz	60–70 Hz		
Conduction velocity				
Axonal	50–80 m/s	58–106 m/s		
Membranal	2.5 m/s	5.4 m/s		
Immunohistochemistry	Slow	Fast		
Reaction with anti-slow MHC IgG	Positive	Negative		
Reaction with anti-fast MHC IgG	Negative	Positive		
	Slow	Fast		
Enzymhistochemistry	I	IIA	IIB (IIX)	
The ATPase activity of the different				
MHCs (I/b, IIA, IIB, IIX) differs in its acid and alkali stability:				
staining after preincubation at pH 10.6	Weak	Strong	Strong	
staining after preincubation at pH 4.6	Strong	Weak	Moderate	
Cytophotometry	SO	FOG II	FOG I	FG
Enzyme activities of muscle fibres				
ATPase	Low	High	High	High
GPDH	Low	Moderate or low	Moderate	High
SDH	Moderate or high	High	Moderate	Low
Morphology	Slow	Fast		
	I	IIA	IIB	
Mitochondria Density	High	High	Low	
Capillarity	High	High	Low	
Size of motoneurons	Small	Large		
Diameter of axon	8–14 μm	9–18 μm		

SO, slow-oxidative; FOG, fast-oxidative glycolytic; FG, fast-glycolytic.

The numerical values are valid for leg muscles of mammals and were taken from Peter et al. (1972); Pette and Vroba (1985); Ebert and Asmussen (1986) and Billeter and Hoppeler (1992).

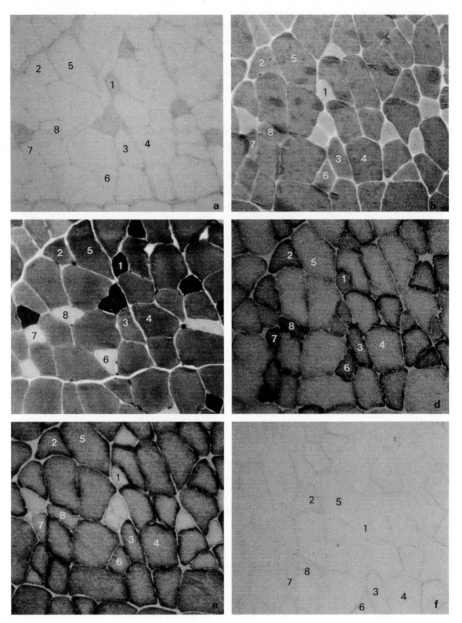

Fig. 2. 10-μm serial cross sections of rat extensor digitorum longus muscle with several histochemical reactions: immunostaining of slow fibres with anti-slow MHC IgG (**a**), ATPase activity after alkalic preincubation at pH 10.4 (**b**), ATPase activity after acid preincubation at pH 4.6 (**c**), succinate dehydrogenase activity (**d**), glycerol-3-phosphate dehydrogenase activity (**e**), and NO-synthase I-associated diaphorase activity ×105. The same 8 fibres are marked and typed in each section: fibre 1=I=SO; fibres 2, 3=IIB=FOG I; 4, 5=IIB=FG; fibres 6, 7, 8=IIA=FOG II. (For fibre typing see Table 1 and Sect. 4.3). Note the metabolic heterogeneity of IIB fibres 2, 3, 4, 5. (**d**) The NO-synthase I-associated diaphorase activity is fibre type-specific demonstrated on the sarcolemma (**f**): the sarcolemma of FOG I fibres (fibres 2, 3) show diaphorase activity, whereas the sarcolemma of SO fibres (fibre 1) is negative

11

IIA, and IIB (including IIX). It was pointed out (Schiaffino et al. 1990) that MHC IIB and IIX cannot be distinguished with ATPase activity (see also Chap. 10). Guth and Samaha (1969) developed the histochemical technique for the myofibrillic ATPase after either acid or alkali preincubation of the histological sections. In this way, after alkali preincubation, weak-stained fibres are type I, dark-stained fibres are type II (Fig. 2b). This corresponds to the slow-fast-typing; type I fibres are slow; type II fibres are fast fibres. After acid preincubation, type I fibres show a dark staining, type II fibres are differentiated into negative IIA fibres and moderatly stained IIB fibres (Fig. 2c). The metabolic properties of these MHC-based fibre types are either equal or different (Fig.2d,e). Our and other studies (Guth and Yellin 1971; Nemeth et al. 1979; Reichmann and Pette 1984; Schiaffino 1990; Larsson 1992) showed that the fibres typed with ATPase reaction are not nesessarily typed metabolically but showed a wide range of metabolic enzyme activities. Overlappings between the fibre types were mainly observed in the oxidative metabolism. That also means that muscle fibres cannot clearly be typed by oxidative enzyme activities only, such as succinate dehydrogenase activity. Moreover, it was shown that during development and in diseased muscles, the pattern of contractile protein isoforms and enzyme activities does not correlate with the ATPase-fibre types (Guth 1973; Guth and Samaha 1972). Detailed studies about the different contractile proteins and enzymes of skeletal muscle are summarized by Pette (1990a).

Two classification systems for muscle fibres, one after physiological, one after meta-bolical properties, were combined for the first time by Peter et al. (1972). The authors measured oxidative and glycolytic enzyme activities in the homogenate of muscles, known as slow or fast twitch muscles. The fibres of these homogeneously composed muscles were designated as SO (slow-oxidative), FOG (fast-oxidative glycolytic) and FG (fast-glycolytic), according to the amount of biochemically measured enzyme activities.

Such studies mentioned above reveal the main problem of muscle fibre analysis: True, it is possible to classify muscle fibres as slow and fast or I, IIA, or IIB in the histological section by qualitative histochemical methods, but their metabolic charac-terization is not possible. In contrast to that, biochemical measurements in the homogenate of muscles allow a metabolic characterization of fibres as long as the muscles concerned consist of one fibre type exclusively. However, apart from few exceptions (i.e. m. tensor fascia latae of the rat), skeletal muscles consist of different fibre types. Studies done so far dealing with the classification of muscle fibres of skeletal muscles which consist of different fibre types were either carried out with qualitative histochemical, biochemical or physiological methods. That means that for methodical reasons they do not give any information about the metabolic situation of a certain fibre type. On the one hand, qualitative histochemistry is performed on single fibres and is able to localize, but it does not give quantitative data. On the other hand, both biochemical and physiological studies inevitably investigate pools of fibres. Consequently, only average data are available. Apart from this, during biochemical investigations the micro-milieu, which contains the molecules including the enzyme molecules that are to be investigated, is destroyed for methodical reasons. As a result, the characteristics of the relevant molecules, i.e. the maximum reaction speed of enzyme molecules V_{max}, are changed, and interactions between them and the cyto-skeleton are no longer possible. Here, cytophotometry as a method of quantitative histochemistry can lead us out of the dilemma. The cytophotometrical method combines the advantages of localizing histochemistry and measuring biochemistry. Therefore, cytophotometry is a useful tool in fibre typing. Detailed information about cytophotometrical fibre typing is given in Chap. 4.

4 Cytophotometry As a Tool in Fibre Typing

4.1
The Basic Principle

Cytophotometry is used as a technique of quantitative enzyme histochemistry based on absorbance measurements of the coloured final reaction product of an enzyme reaction in the histological section (for review, see Van Noorden and Butcher 1991). The main advantage of the cytophotometrical method is to enable measurements in defined microscopical structures such as muscle fibres (Pai et al. 1982; Reichmann and Pette 1982,1984; van der Laarse et al. 1984; Vetter et al. 1984; Blanco et al. 1988; van der Laarse et al. 1989; Blanco and Sieck 1992, this study). The levels of metabolic enzyme activities which are demonstrated histochemically by their reaction product vary in the different muscle fibres. The absorbance of the reaction product can be assumed to be proportional with its concentration in the histological section (Krug 1980). Moreover, fulfilling a set of criteria for reliability and validity of the enzyme reaction (Stoward,1980, van Noorden and Frederiks 1992), the absorbance of the reaction product is additionally proportional with enzyme activity and can serve as a measure for that. Direct comparison of cytophotometry and biochemistry is possible by calibration of cytophotometrical values (van Noorden and Frederiks 1992). In this way, data of both methods can be expressed as moles of substrate converted per unit time per unit wet weight of tissue. However, for many investigations such as muscle fibre analysis, absolute values of enzyme activities are not necessary. It is rather the comparison of enzyme activities in different structures or under changing experimental conditions that is of interest. In these cases, one can use enzyme activities as percentages related to a reference value, and calibration procedures are unnecessary.

4.2
The Enzymes Which Are Useful for Cytophotometrical Fibre Typing

In which way is cytophotometry used as a tool in fibre typing? The basic idea for the use of cytophotometry in fibre typing is to measure activities of such enzymes in muscle fibres which give information about contractility and oxidative and glycolytical metabolism. Such enzyme activities are known to vary in different fibre types. Furthermore, for practical use, enzyme activities for cytophotometrical analysis should be quantifiable as simply and reliably as possible. According to these ideas, the activities of the following enzymes are used for cytophotometrical fibre typing: myofibrillic adenosine triphosphatase (E.C. 3.6.1.32, ATPase) as a marker of contractility, glycerol-3-phosphate dehydrogenase (E.C. 1.1.99.5, GPDH) as a marker of glycolysis and succinate dehydrogenase (E.C. 1.3.5.1, SDH) as a marker of oxidative metabolism. These enzymes are structure bound proteins which are localized at the inner mitochondrial membrane and in the case of ATPase at the myofilaments.

The reasons for the ability of the chosen enzymes to characterize the muscle fibre properties mentioned above are as follows:

1. Adenosine triphosphatase (ATPase)

   The contraction activity of myofibrils depends on the ATPase activity of myosin or actomyosin complex, respectively. The myosin ATPase in skeletal muscle splits the terminal phosphate bond of ATP whereby energy is released. Using histochemical methods, several ATPase systems are included: myofibrillic contractile ATPase systems, the oxidation-phosphorylation coupling mitochondrial ATPase, and the calcium-transporting ATPase of the sarcoplasmic reticulum (Borgers et al. 1991). With one histochemical method several ATPase activities could be demonstrated simultaneously. For muscle fibre typing, the traditional demonstration of the so-called myofibrillic ATPase activity is useful according to Padykula and Herman (1955).. It includes the activities of ATPase located at the myosin molecule as well as the ATPase located at the inner mitochondrial membrane. It was shown (van der Laarse et al. 1986) that the activity of myofibrillic ATPase strongly correlates with the contraction time of muscle fibres. Consequently, the myofibrillic ATPase activity is a suitable measure for contractility.

2. Glycerol-3-phosphate dehydrogenase (GPDH)

   The GPDH which is located in the mitochondria plays a key role in the glycerol phosphate shuttle from mitochondrium to the the glycolytical system in the cytosol. The enzyme catalyzes the following reaction (Stoward et al. 1991): Glycerol-3-phosphate + acceptor ⊗ Dihydroxyacetone phosphate + reduced acceptor. The cytosolic component of the shuttle system is the soluble NAD+- dependent GPDH. It was shown for several muscles of mammals that the activity of mitochondrial GPDH is proportional to the activity of glycolytical key enzymes (Pette and Bücher 1963; Pette 1966; Pieper et al. 1984a). Therefore, it is justified to use the well localizable mitochondrial GPDH activity as an indirect marker of the glycolytical activity (Reichmann and Pette 1984; Schmalbruch 1985a; van Noorden and Butcher 1991; Billeter and Hoppeler 1992).

3. Succinate dehydrogenase (SDH)

   This mitochondrial-bound enzyme is a frame of reference both in the citric acid cycle and in the respiration chain (Stoward et al. 1991). It catalyzes the conversion of succinate to fumarate and was already succesfully used as a measure of oxidative capacity in skeletal muscles.

4.3
The Physiological–Metabolic Muscle Fibre Typing By Cytophotometry

The cytophotometrical measurements of the activities of myofibrillic ATPase, GPDH and SDH in cryostat sections from skeletal muscles enable the physiological and

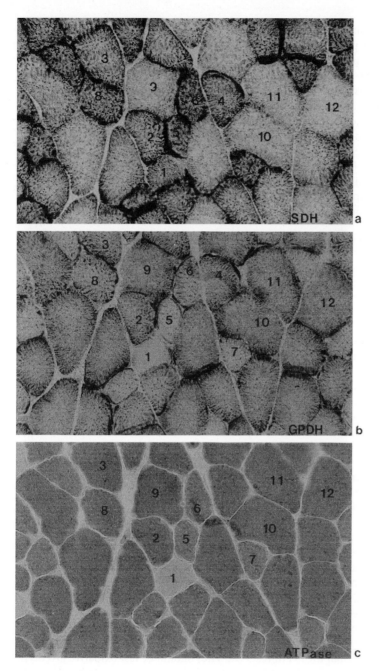

Fig. 3. 10-µm serial cross sections of rat extensor digitorum longus muscle, reacted for succinate dehydrogenase (*SDH*; **a**), glycerol-3-phosphatedehydrogenase (*GPDH*; **b**) and adenosine triphosphatase (ATPase; **c**) ×260. The same 12 fibres are identified with different enzyme activities

metabolic characterization of muscle fibres. This means that the physiological–metabolic fibre typing is possible not only in homogeneously but also in heterogeneously composed skeletal muscles.

The procedure of physiological–metabolic fibre typing in cryostat sections of skeletal muscles is as follows: The activities of the three characteristic enzymes have to demonstrate histochemically in three serial cryostat sections (Fig. 3). For this, usual histochemical techniques (Lojda et al. 1976) have been used. In each fibre the activities of SDH (Fig. 3a) and GPDH (Fig. 3b)as well as ATPase (Fig. 3c)are demonstrated by the coloured final reaction products. According to the remarks made in Sect. 4.2, it should now be possible to obtain information about the intensity of oxidative and glycolytic metabolism as well as contractility in the same fibre. The cytophotometrical measurements of the three enzyme activities in one and the same fibre lead to the following classification of the muscle fibres:

Fibre type	Physiological	Metabolic properties
SO	Slow	Oxidative
FOG	Fast	Oxidative glycolytic
FG	Fast	Glycolytic

As an example, in the following this fibre typing is demonstrated in detail for the rat extensor digitorum longus muscle. Twelve fibres are marked in serial sections reacted for the respective enzyme (Fig. 4). Whereas the ATPase picture reflects the well-known characteristic of EDL as a fast twitch muscle consisting of 95% fast (dark stained) and 5% slow (light stained) fibres, SDH and GPDH pictures are more heterogeneous. It will be shown that each fibre is characterized by the cytophotometrical data of the three enzyme activities: ATPase, SDH, and GPDH. For this, the cytophotometrical data about enzyme activities of the 12 marked fibres are demonstrated in Fig. 4. The interpretation of graphs is as follows: fibre 1 shows low ATPase activity, typing the fibre as type I and identical to the type "slow" (see also Table 1). At the same time, fibre 1 has the smallest glycolytic (GPDH) and a moderate oxidative (SDH) activity, compared to the other investigated fibres. Consequently, fibre 1 is typed as a slow-oxidative fibre, SO. Here, only one SO fibre was detected because of the small population of SO fibres at 5% in the EDL. The analysis of a great number of muscle sections confirmed the findings from this example. The fibres 2–12 show high ATPase activity; they are of type II identical to the type "fast" (see also Table 1). The SDH and GPDH activities vary within the fast fibres and subdivide the fast fibres according their metabolic equipment. Comparing the cytophotometrical data of the evaluated fibres, the fast fibres 9–12 have the smallest SDH activity combined with the highest GPDH activity. Therefore, these fibres are fast glycolytic fibres, FG. The fast fibres 2–8 on the one hand show higher SDH and on the other hand lower GPDH activity than the FG fibres 9–12. They should be fast oxidative glycolytic fibres, FOG. One can even see on the muscle sections (Fig. 3a,b) what the

---

All cytophotometrical measurements which form the basis of this book were performed with the microscope photometer MPM 200 with a scanning table (Carl Zeiss, Oberkochen, Germany).

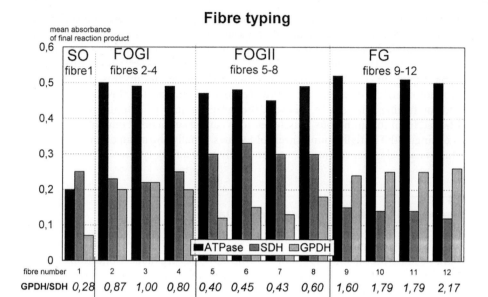

Fig. 4. Cytophotometrical values (mean absorbance of final reaction product) of the enzyme activities and the GPDH/SDH activity quotient of the 12 marked fibres from Fig. 3. The fibres were typed into slow-oxidative (*SO*), fast-oxidative glycolytic (*FOG I, II*) and fast-glycolytic (*FG*)

cytophotometrical data reveal: the FOG fibres differ from each another in their metabolic capacity. Both oxidative (SDH) and glycolytic (GPDH) activities show conspicuous heterogeneity within the FOG fibres. In this respect it is interesting to compare the levels of SDH activity of the different FOG fibres with that of SO fibres. Whereas some FOG fibres (5–8) show higher SDH activity than the SO fibre, other FOG fibres (2–4) have a similar level of SDH activity to that of SO fibre. Knowing that SO and FOG fibres belong to different ATPase fibre types this confirms the findings of Reichmann and Pette (1982). These authors revealed that ATPase fibre types show overlapping oxidative activity. The cytophotometrical data prove clearly that it is not possible to differentiate SO and FOG fibres by the SDH reaction only. But, if one includes the two enzymes SDH and GPDH, the differentiation of SO and FOG fibres becomes possible. For this, the activity quotient GPDH/SDH gives numerical values which differ from fibre type to fibre type, see Fig. 4. In this way, the GPDH/SDH activity quotient can serve as a discriminator for fibre typing. Now, the subdivision of FOG fibres into 2 types, FOG I and FOG II, emerges. FOG I fibres are characterized by moderate SDH and moderate GPDH activity; FOG II fibres show high SDH and low or moderate GPDH activity.

Moreover, changes of the GPDH/SDH activity quotient of a fibre type characterize changes in the metabolic profile of this fibre type which can occur under different experimental conditions. Such investigations are presented in Chap. 5.

It should be mentioned that in practical fibre typing and to obtain results about changing enzyme activities, much more than 12 fibres must be evaluated. Often thousands of fibres are to be analysed. Then the laborious detection and cytophotometrical measurements of one and the same fibre in serial cross sections are not practicable for all fibres. A routine technique is necessary as described by Punkt et al. (1993). In brief, only 30–35 fibres including the spectrum of diversity are to be typed in serial sections as described above. From this, the range of cytophotometrical data for each fibre type is evident, and the threshold values are used for further analysis. In this way, the large number of cytophotometrical data obtained from the numerous muscle sections can be subdivided into different classes according to the initially found threshold values. Finally, the mean cytophotometrical value of each class represents the enzyme activity of a fibre type.

4.4
The Reliability of Cytophotometrical Data

The reliability of cytophotometrical data as a measure of enzyme activity should be guaranteed before statements about enzyme activities are made. Conditions for valid quantification of enzyme histochemical reactions by means of cytophotometry are often discussed in the literature (for references, see van Noorden and Butcher 1991). A meaningful criterion for the reliability of cytophotometrical data of enzyme activity is the correlation with biochemically determined enzyme activity in the same material. As mentioned in Sect. 4.1, cytophotometrical data of enzyme activities are determined in microscopically defined areas of homogeneous and heterogeneous tissues. On the other hand, biochemical measurements of enzyme activity always reflect overall activity in tissue homogenates. In case of homogeneous enzyme reaction in the muscle section it is possible to compare directly cytophotometrical data from each site of the section with biochemically measured enzyme activity in homogenate. Consequently, to compare data from both methods, tissue should be used in which enzyme activity is homogeneously distributed. For this purpose, myocardium of the left ventricle of rats is a suitable tissue, because the final reaction product of several enzymes is homogeneously distributed. In case of a close correlation between cytophotometrically and biochemically determined enzyme activity, values obtained by cytophotometry must be reliable measures of enzyme activity in heterogeneous tissues as well, such as skeletal muscle. We have shown a strong correlation between cytophotometrical and biochemical data for ATPase, SDH and GPDH activity (Punkt et al. 1984, 1986, 1997). The graphs in Figs. 5 and 6 demonstrate the correlation for SDH and GPDH established by linear regression analysis. We measured biochemically the activity of the glycolytic key enzyme lactate dehydrogenase (LDH) for correlation with cytophotometrically determined GPDH activity. The reason for that is as follows: As already shown by Pette and Dölken (1975), the activity of GPDH in heart muscle is up to 350-fold lower than that of LDH. A change of the already very low activity by 15%–40% as observed cytophotometrically in different experimental groups would be within the biochemical test variance. As mentioned above, the strong correlation between mitochondrial GPDH and total LDH activity justifies the use of GPDH as a marker of glycolytic metabolism in cytophotometry. In summary, the strong correlation between cytophotometrical data and activity of the respective enzymes as measured biochemically justifies the use of cytophotometry in quantitative enzyme histochemistry.

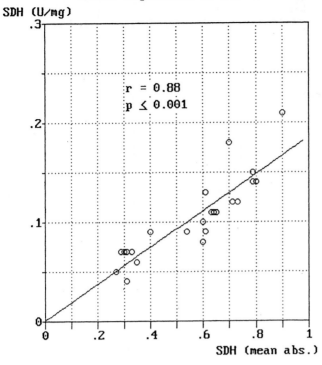

Fig. 5. Linear regression of cytophotometrical (*abscissa*; mean absorbance of reaction product) and biochemical (*ordinate*) data of succinate dehydrogenase (*SDH*) activity in the myocardium of rats from different experimental groups. *r*, correlation coefficient; *p*, probability of error

There are several traps which can make the cytophotometrical data unreliable. Which restrictions are to be accepted to obtain reliable cytophotometrical data correlating with enzyme activities? Firstly, only unfixed tissue should be used for cytophotometrical measurements. Especially aldehyde fixation inhibits enzyme activities and unreliable cytophotometrical measures are to be expected. Additionally, in the case of block-fixation, the penetration of the fixative through the tissue sample is usually reduced along the way from the superficial to the deep part of the tissue block. Consequently, the fixation effect on enzyme activities varies within the tissue block and on the cryostat sections. This was shown in the rat myocardium by Punkt et al. (1987). Secondly, to compare tissue samples from different experimental groups, methodical variations which are caused e.g. by section thickness and incubation procedure must be excluded. For this, the samples to be compared should be mounted on one and the same cryostat desk and cut together in a one-section step. In this way, the cryostat sections of the different samples have the same thickness and are located on the same slide. Consequently, the histochemical conditions to which they are subjected in the further procedure are identical for these sections. For comparison of muscle sections on different slides, reference values can be used (Punkt et al. 1984,

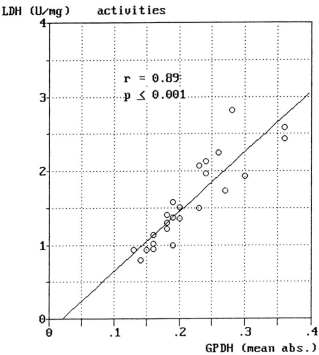

Fig. 6. Linear regression of cytophotometrical (*abscissa*; mean absorbance of reaction product) and biochemical (*ordinate*) data of glycerol-3-phosphate dehydrogenase (*GPDH*) and total lactate dehydrogenase (*LDH*) activity in the myocardium of rats from different experimental groups. *r*, correlation coefficient; *p*, probability of error

1985). Another way to eliminate differences of absorbance caused by the variation of section thickness between sections from different slides is to set the absorbance of the background (unreduced NBT, tissue components) to 0.

Finally, the apparative parameters for cytophotometrical measurements are to be chosen in such a way that the distributional error is avoided as much as possible. That means for scanning cytophotometry with a microscope photometer that the measuring spots along a scanning line in the histological section do not overlap each other, or in other words that the step size of the scan must be larger than the diameter of the measuring spot.

The mentioned restrictions for reliable cytophotometrical data can be fulfilled easily and should not be a barrier to apply the cytophotometry in muscle analysis.

The purpose and advantage of cytophotometry in muscle analysis would not be fully used if cytophotometry was limited to being a tool in fibre typing only. Cytophotometrical fibre typing is always a means to an end. It is rather the use in investigations of changing fibre metabolism under changing experimental conditions which is the main advantage of cytophotometrical muscle analysis.

5 Changes of Muscle Fibre Properties Under Physiological and Pathological Conditions

Biochemical measurements have shown that the metabolism of the whole muscle changes under varying experimental conditions (Simoneau 1990; Schantz et al. 1997; Taylor and Bachman 1999; Prior et al. 2001). Moreover, it was observed that the metabolic properties of the muscle fibre changed earlier and more quickly than its contractility (Hoppeler 1990). However, biochemical measurements do not give any information about the extent of the metabolic changes in a certain fibre type or about the share of a certain fibre type of the measured activity changes of the whole muscle respectively. At this point, cytophotometrical measurements of enzyme activities in the metabolically characterized fibres are necessary. The enzymes which were suitable markers for fibre typing (Sect. 4.2.) are at the same time useful for characterizing metabolic changes of the fibre under changing conditions. Changes of the activity quotient GPDH/SDH of a fibre type characterize changes of the metabolic profile of this fibre type. This means that cytophotometrical measurements of the three chosen enzyme activities in muscle sections from different experimental groups simultaneously provide fibre typing in SO, FOG and FG fibres and changes of the metabolic profile in these fibre types.

Apart from enzyme activities, also other fibre properties such as population and diameter change under altered conditions. The next paragraphs deal with changes in fibre properties under several physiological and pathological conditions. Summarizing all effects on a fibre type, conclusions about the adaptability of this fibre type become possible.

5.1
Regional Differences

5.1.1
Illustration of the Problem

The cytophotometrical method enables the typing of muscle fibres as well as fibre type related measurements of changes in fibre properties under different conditions. To interpret the measured changes clearly, we will first answer the following question: Do those fibres which belong to a certain fibre type of a given muscle all have the same properties, or do they vary depending on the location of the fibre in the muscle? If the latter is true, such regional differences have to be taken into consideration when investigating changes in fibre properties due to ageing, training, diseases or other factors.

Several studies deal with this problem, evoking doubts about the uniformity of the motor unit. As already mentioned in the Chap. 1, a motor unit (the α-motoneuron and all of its innerved muscle fibres) comprises fibres of the same physiological type. Cross-innervation and chronic stimulation studies suggest that muscle fibre heterogeneity in skeletal muscle is primarily related to the properties of the innervating motoneuron (e.g. Pette and Vroba 1985), which determine the fibre type physiologically. However, the motor unit is not uniform! Variations in the activities of enzymes involved in energy metabolism have been observed within single motor units (Kugelberg and Lindegren 1979; Martin et al. 1988a,b; Larsson 1992), suggesting that factors other than motoneuron properties may determine skeletal muscle fibre heterogeneity. One factor could be variation of fibre type properties on the basis of the location of the fibre within the muscle. Contradictory reports on fibre size related to depth of sampling site describing differences along the superficial-deep axis of muscle have also been published (Polgar et al. 1973; Halkajaer-Kristensen and Ingeman-Hansen 1981; Henriksson-Larsen et al. 1985; Larsson 1992). Moreover, large variations in relationships between fibre size and enzyme activities or physical activity have been reported (Costill et al. 1976a,b; Prince et al. 1976; Bylund et al. 1977; Jansson et al. 1978, Schantz et al. 1983; Larsson 1992). These variations could be due to regional differences in metabolic and morphological properties of muscle fibres which are not only related to the transversal (superficial-deep) axis but also to the longitudinal axis from the origin to the insertion of the muscle.

In the following, we will show that the muscle fibres of a given fibre type, that means of a given motor unit, vary in their morphological and metabolic properties dependent on their location within the muscle. Seen from this angle, the motric unit is to be considered as nonuniform.

5.1.2
Experimental Basis

To show the regional differences of fibre properties within the muscle, the results from an extensive cytophotometric-morphometrical analysis of two rat skeletal muscles are demonstrated. The extensor digitorum longus (EDL) and soleus (SOL) muscles were chosen for the topographical analysis, because they are often used for investigations of physiological or pathological effects. Therefore, topographical variations within these muscles should be of interest. The muscles from 28-month-old rats were subdivided into defined muscle regions for quantitative analysis (Fig. 7). Along the longitudinal axis the muscle was dissected into three samples from origin, middle part and insertion. Along three transversal axes, which are located near the origin, in the middle part and near the insertion, superficial, central and deep regions were differentiated.

The muscle samples of this experiment as well as of all other experiments which are described in Chap. 5 were treated according to the following general pattern: They were frozen in liquid nitrogen, cut into 10-μm sections and incubated for the respective enzyme activities, SDH, GPDH and myofibrillic ATPase (see Sect. 4.2).

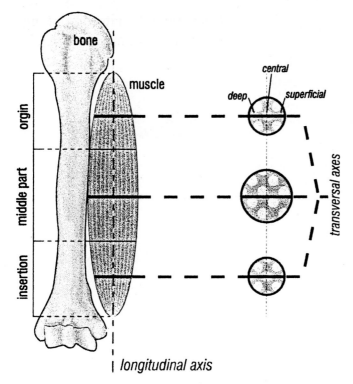

Fig. 7. Subdivision of the muscle into defined muscle regions for quantitative analysis. Along the longitudinal axis the muscle was dissected into three pieces from origin, middle part and insertion. Along three transversal axes, which are located near the origin, in the middle part and near the insertion, superficial, central and deep regions were differentiated

End-point cytophotometry was performed, the fibres were typed, and the mean relative enzyme activities of the fibre types SO, FOG and FG were calculated, for more information see also Sect. 4.3. The fibre type distribution and the fibre cross areas in both muscles were estimated with a videoplan system (Carl Zeiss).

5.1.3
Changes of Fibre Properties Along the Longitudinal Axis from the Origin to the Insertion of the Muscle

5.1.3.1
Fibre Type Distribution

The fibre type distributions of EDL and SOL muscles along the longitudinal axis of the muscle are demonstrated in Figs. 8 and 9.

Fig. 8. Topographical variations in fibre type distribution in extensor digitorum longus muscle *(EDL)*. Fibre types: *SO*, slow-oxidative; *FOG*, fast-oxidative glycolytic; *FG*, fast-glycolytic. Differences between the proportion of FG fibres between the origin and middle part and between the middle part and insertion of muscle were shown to be significant ($p<0.05$ for origin and $p<0.01$ for insertion) by *t*-test

Fibre type distribution along the longitudinal axis of EDL

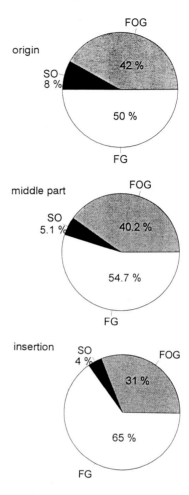

The EDL muscle (Fig. 8) consists mainly of FG and FOG fibres and a small number of SO fibres. The characteristic composition of muscle was observed in all regions investigated, but variations of fibre type composition were found along the axis from the origin to the insertion. The population of FG fibres increases by 15% from the origin to the insertion at the expense of the FOG fibres (decreasing by 11%) and SO fibres (decreasing by 4%).

The characteristic fibre type of SOL muscle (Fig. 9) is SO, whose contribution increases from the origin (60%) to the insertion (90%). This increase is accompanied by a decrease in the percentage of FG fibres. At the origin, 30% of the fibres are FG fibres whereas only 1% of the fibres are FG fibres at the insertion.

Fig. 9. Topographical variations in fibre type distribution in soleus muscle *(SOL)*. Fibre types: *SO*, slow-oxidative; *FOG*, fast-oxidative glycolytic; *FG*, fast-glycolytic. Differences between the proportion of SO and FG fibres between the origin and middle part and between the origin and insertion of muscle were shown to be significant (*p*<0.01) by *t*-test

Fibre type distribution along the longitudinal axis of SOL

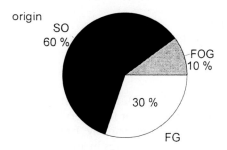
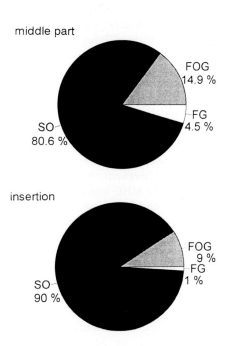

An explanation for these changes in fibre type distribution may be as follows: Muscle fibres insert in a staggered fashion; they form a parallelogram together with the tendon sheets at which they more or less obliquely insert (Schmalbruch 1985b). That does not mean that all fibres run from tendon to tendon. Fibres can terminate within the muscle belly. In pennate or bi-pennate muscles such as EDL and SOL muscles, some of the parallel arranged fibres in the middle part of muscle insert earlier at the tendon than others. Only some fibres extend over the whole muscle length (Tittel 1981). In this way, fibre number and distribution varies within the muscle along the longitudinal axis from the origin to the insertion. The results shown above suggest that in EDL, FG fibres (origin) or FOG fibres (insertion) insert earlier and in SOL SO fibres (origin) or FG fibres (insertion) insert earlier. Therefore, the ratio of that fibre type which is typical for the considered muscle (FG for EDL muscle, SO for SOL muscle) is highest at the insertion.

5.1.3.2
Fibre Type Related Enzyme Activities

Enzyme activities of the fibre types changed along the longitudinal axis in different ways.

EDL Muscle (Fig. 10). In contrast to SDH activity, which shows only a significant decrease in FG fibres at the insertion (Fig. 10b), the GPDH activity of FOG and FG fibres increases continuously from the origin to the insertion (Fig. 10a). The GPDH/SDH ratio increases in all fibre types of EDL from the origin to the insertion (Table 2). Moreover, the GPDH/SDH activity ratio of FOG fibres is more similar to that of SO fibres than that of FG fibres. Like the GPDH activity, the ATPase activity of type I (slow) and type II (fast) fibres increases (Fig. 10c).

SOL Muscle (Fig. 11). In contrast to EDL muscle, especially the SDH activity changes in all fibre types along the longitudinal axis with the highest activity in the origin and the lowest activity in the insertion part (Fig. 11b). GPDH activity shows no significant differences in the different muscle regions (Fig. 11a). Consequently, an increased GPDH/SDH activity ratio was found in the region of the insertion (Table 2). ATPase activity of all fibres is highest at the insertion (Fig. 11c).

These results show variations in enzyme activities of a given fibre type along the longitudinal muscle axis. Here, the homogeneous distribution of enzyme activities over the length of fibres comes into discussion. It is known (Staron et al. 1987) that the distribution of the MHC isoform along the fibre length is nonuniform even in the skeletal muscle under normal conditions, and the nonuniformity increases under muscle fibre transformation. The distribution of MHC mRNA along the fibre length is also reported to be nonuniform (Peuker and Pette 1997). Up until now, this fact is not fully understood, and many questions remain, e.g. whether a single nucleus of the multinuclear muscle cell produces mRNA of a single isoform or of multiple isoforms. One can assume that the mRNA of other proteins, such as GPDH and SDH, also varies along the fibre length, resulting in a variation of distribution and activitiy of these enzymes.

If GPDH is considered as a marker for glycolytic metabolism (Pieper et al. 1984a; Schmalbruch 1985a; van Noorden and Butcher 1991; Punkt 1997) and SDH as a marker of oxidative metabolism (van der Laarse et al. 1989), then the following can be stated:

1. The fibres of EDL become more glycolytic from the origin to the insertion expressed by an increase of GPDH/SDH activity quotient.
2. The fibres of SOL muscle become less oxidative from the origin to the insertion, which also results in an increase of GPDH/SDH activity quotient.
3. Glycolytic activity changes mainly in the fast fibre type (FOG and FG) along the longitudinal axis of a muscle which is characterized by predominantly glycolytic metabolism such as EDL muscle.
4. The oxidative capacity changes significantly in all fibre types along the longitudinal axis of a muscle with predominantly oxidative metabolism, such as SOL muscle.

Changes along the longitudinal axis of EDL

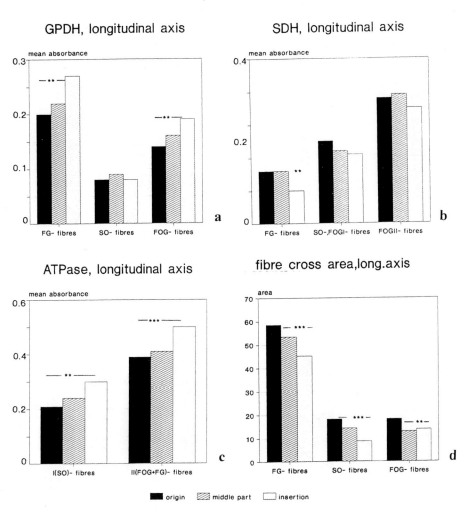

Fig. 10. Fibre type specific changes in activities of glycerol-3-phosphate dehydrogenase *(GPDH)*, succinate dehydrogenase *(SDH)* and myofibrillic adenosine triphosphatase *(ATPase)* and changes of fibre cross area (area, given in $\mu m2 \times 10-2$) along the longitudinal axis from the origin to the insertion of extensor digitorum longus muscle *(EDL)*. Fibre types: *SO*, slow-oxidative; *FOG*, fast-oxidative glycolytic; *FG*, fast-glycolytic. Differences between origin and insertion were analysed with the *t*-test; **$p<0.01$, ***$p<0.001$

Accepting the ATPase as a marker of contractility (van der Laarse et al.; 1986), the increased ATPase activity at the insertion of both muscles may be interpreted as an indication of increased contractility in this region.

Changes along the longitudinal axis of SOL

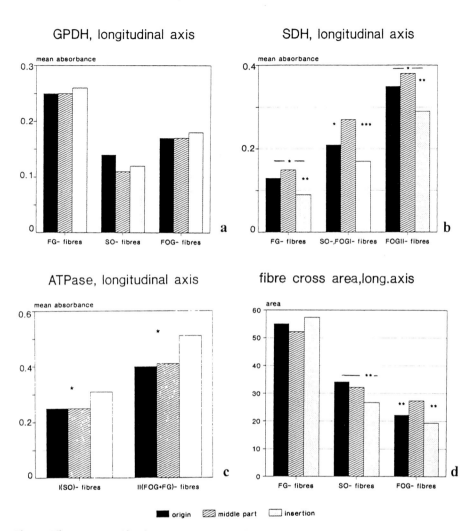

GPDH, longitudinal axis

SDH, longitudinal axis

ATPase, longitudinal axis

fibre cross area,long.axis

■ origin   ▨ middle part   ☐ insertion

Fig. 11. Fibre type specific changes in activities of glycerol-3-phosphate dehydrogenase *(GPDH)*, succinate dehydrogenase *(SDH)* and myofibrillic adenosine triphosphatase *(ATPase)* and changes of fibre cross area (area, given in $\mu m2 \times 10^{-2}$) along the longitudinal axis from the origin to the insertion of soleus muscle *(SOL)*. Fibre types: *SO*, slow-oxidative; *FOG*, fast-oxidative glycolytic; *FG*, fast-glycolytic. Differences between origin and insertion were analysed with the *t*-test;*$p<0.05$, **$p<0.01$, ***$p<0.001$

The results shown above demonstrate: on the one hand the fibre type distribution, and on the other hand metabolic profiles of fibres are adapted to topographically different functional demands. The fibres of EDL muscle (a fast-force muscle) shows the highest contractile and glycolytic capacities near its insertion, where the highest percentage of FG fibres is found. At the insertion, the strongest effects of muscle work are expected. The fibres of SOL muscle (a muscle of endurance) shows highest oxidative capacity in the middle part of muscle, where 80% SO fibres are found, suggesting a main role in endurance function.

5.1.3.4
Fibre Cross Areas

While the fibre cross areas of all fibre types of EDL muscle decrease along the longitudinal axis from the origin to the insertion (Fig. 10d), in SOL muscle the cross areas of the different fibre types change differently (Fig. 11d). The SO fibres decrease, while the FOG fibres show the largest cross area in the middle part of muscle, and the FG fibres do not change significantly in cross areas.

5.1.4
Changes of Fibre Properties Along the Transversal Axis from the Superficial to the Deep Region of Muscle

5.1.4.1
Fibre Type Distribution

The heterogeneous distribution pattern of fibre types over muscle cross sections is shown in Fig. 12. In the investigated muscles, SDH-positive fibres, SO and FOG, are located more deeply than superficially. This is reasonable, because more oxidative fibres can realize the higher muscle tonus better, which is necessary near the bone.

5.1.4.2
Fibre Type Related Enzyme Activities

Wide topographical variations in enzyme activities of muscle fibres from a given type were found for both muscle investigated from the superficial to the deep region. The detailed data cannot be demonstrated here, but a few general results will be considered.

EDL muscle (here not demonstrated): GPDH activity of all fibre types of EDL muscle increases along the transversal axis from the superficial to the deep region. In contrast to GPDH activity, changes in SDH activity along the transversal axis are dependent on the location of the axis in the muscle. Moreover, the SDH activity changes depending on the fibre type. The changes in the GPDH/SDH ratio are listed in Table 2.

SOL muscle (Fig. 13): Changes of SDH activity along the transversal axis are independent of the location of the axis within the muscle. A continuous increase of SDH activity was measured in all fibre types from superficial to deep location (Fig. 13a). Similarly to SDH, but to a lower extent, GPDH activity increases in all fibre types continuously along the superficial-deep axis (Fig. 13b). This trend is observed along

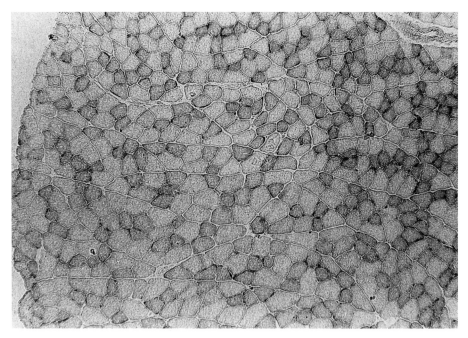

Fig. 12. 10-μm cross section of soleus muscle near the origin reacted for succinate dehydrogenase *(SDH)*, ×6.3. From superficial *(left side)* to deep *(right side)* the number of SDH positive fibres increased

all transversal axes investigated. Consequently, the GPDH/SDH ratio decreases in all fibre types of SOL along the three transversal axes (Table 2).

Generally, in both EDL and SOL muscles, the oxidative capacity of all fibre types becomes higher along the superficial-deep axis located in the middle part of muscles. Both slow and fast fibres show this trend, suggesting that such regional differences in oxidative metabolism are not dependent on the fatigue resistance of the fibre type. More oxidative capacity of the fibres in the depth on the one hand and increased number of oxidative fibres in the depth on the other hand, as shown above, are necessary for more tonus work near the bone. Additionally, an increased glycolytic capacity was found in all fibre types. Changes in the GPDH/SDH ratio are interpreted as metabolic shift. The decrease of GPDH/SDH ratio from superficial to deep in all fibre types of SOL muscle indicates a shift to a more oxidative metabolism (Table 2). In contrast, the fast twitch EDL muscle with predominantly glycolytic metabolism shows this shift to a more oxidative metabolism along the superficial-deep axis only in the glycolytic fibres FG.

Moreover, the ATPase activity decreases mainly in type II (fast) fibres of both muscles along the superficial-deep axis (demonstrated for SOL in Fig. 13c). This indicates that the deep fast fibres have lower capacity for contractility than the superficial ones. In conjunction with the shift to more oxidative metabolism one could conclude that the character of fast fibres in deeper layers of muscle becomes more similar to SO fibres.

Table 2. Changes in the GPDH/SDH ratio of the different fibre types along three transversal axes (near the origin, in the middle part and near the insertion) in extensor digitorum longus and soleus muscles

	Superficial	Central	Deep	Significance superficial–deep
EDL				
FG				
Origin	1.05	1.0	0.91	*
Middle Part	1.6	1.28	1.47	*
Insertion	2.8	2.0	2.0	*
SO				
Origin	0.34	0.29	0.37	
Middle Part	0.5	0.4	0.5	
Insertion	0.58	0.62	0.70	*
FOG I				
Origin	0.52	0.47	0.60	*
Middle Part	0.83	0.73	0.92	*
Insertion	1.05	1.0	1.0	
FOG II				
Origin	0.34	0.34	0.34	
Middle Part	0.52	0.49	0.54	
Insertion	0.52	0.70	0.70	*
SOL				
FG				
Origin	2.5	2.08	1.4	**
Middle Part	1.2	1.57	1.0	*
Insertion	3.3	2.44	1.92	**
SO				
Origin	0.6	0.52	0.44	**
Middle Part	0.6	0.32	0.41	**
Insertion	1.28	1.09	0.63	**
FOG I				
Origin	1.0	0.95	0.72	**
Middle Part	0.93	0.58	0.65	**
Insertion	2.0	1.36	1.0	**
FOG II				
Origin	0.64	0.58	0.54	*
Middle Part	0.58	0.58	0.46	*
Insertion	0.96	0.80	0.55	**

Values are means; *$p<0.05$, **$p<0.01$. Abbreviations: EDL, extensor digitorum longus muscle; SOL, soleus muscle; SDH, succinate dehydrogenase; GPDH, glycerol-3-phosphate dehydrogenase; FG, fast-glycolytic fibre type; SO, slow-oxidative fibre type, FOG I, fast-oxidative glycolytic fibre type with moderate SDH activity; FOG II, fast-oxidative glycolytic fibre type with high SDH activity; FG, fast-glycolytic fibre type.

Enzyme activity changes and fibre cross areas along transversal axes of SOL

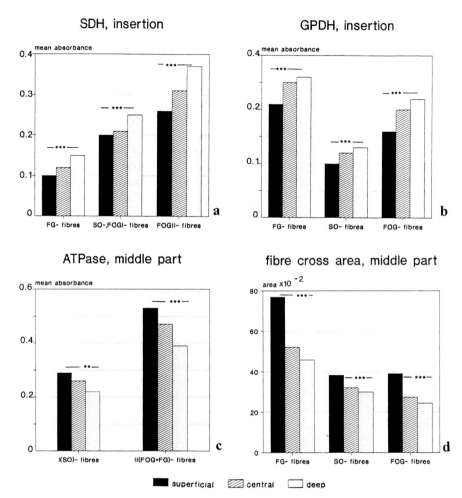

Fig. 13. Fibre type specific changes in activities of glycerol-3-phosphate dehydrogenase *(GPDH)*, succinate dehydrogenase *(SDH)* and myofibrillic adenosine triphosphatase *(ATPase)* and changes of fibre cross area (area, given in µm2) along transversal (superficial-deep) axes at the insertion and in the middle part of soleus muscle *(SOL)*. Fibre types: *SO*, slow-oxidative; *FOG*, fast-oxidative glycolytic; *FG*, fast-glycolytic. Differences between origin and insertion were analysed with the *t*-test;**$p<0.01$, ***$p<0.001$

5.1.4.3
Fibre Cross Areas

The cross areas of the different fibre types change significantly along the transversal axis from the superficial to the deep region in each part of the muscles. Differences between EDL and SOL muscles were observed.

The cross areas of all fibre types of SOL muscle decrease from the superficial to the deep region, independent of the region, shown for the middle part (Fig. 13d). Moreover, SOL fibre cross areas correlate negatively with SDH activity for all fibre types (Fig. 14). These results agree with findings of Sieck et al. (1987) in diaphragm and of Larsson (1992) in tibialis anterior muscle of rats. The functional interpretation of such a relationship may be related to the diffusion distance for metabolic substrates. The smaller the diffusion distance, the better these substrates are utilized and better fatigue resistance of muscle fibres can result.

In contrast to SOL muscle, the change of cross areas of SO and FOG fibres is dependent on the location of the transversal axis in EDL muscle (Fig. 15). No correlations between fibre cross areas and enzyme activities were found. This suggests that in EDL, but also in other muscles, the outcome of measurements depends on the site where the sample is taken from the muscle. For example, the measurements of decreasing fibre cross areas along the superficial-deep axis of rat tibialis anterior muscle in Larsson's study (1992) were made at the greatest girth of the muscle. In contrast to that, Henriksson-Larsen (1985) found an increase in fibre size of all fibres along the superficial-deep axis near the origin in human tibialis anterior muscle. We have shown (Fig. 15a) that near the origin of rat EDL muscle SO and FOG fibres increased along the superficial-deep axis. Contradictory results in the literature may originate from where samples are taken from a muscle.

5.1.5
Conclusions

Muscle fibres can adapt to the demands in different parts of the muscle by varying their size and enzyme activities. This adaptation depends on the physiological and metabolic character of muscle. Moreover, regional variations in fibre size and enzyme activities within a muscle may suggest that adaptation to physical demands need not be uniform throughout the whole muscle. It was indicated that a single region can not be taken as representative for the whole muscle. To investigate changes in muscle fibre properties due to biological or pathological changes, these considerations should be taken into account.

5.2
Development and Ageing

As described in Chap. 2, a plan of muscle fibre diversity is introduced early in the histogenesis. Two distinct types of muscle fibres -slow and fast- were generated. The specification of subtypes occurs later in development. The exact time for this is unknown, it happens probably during the maturation (postnatal development). As the

Correlation of SDH activity and fibre cross area

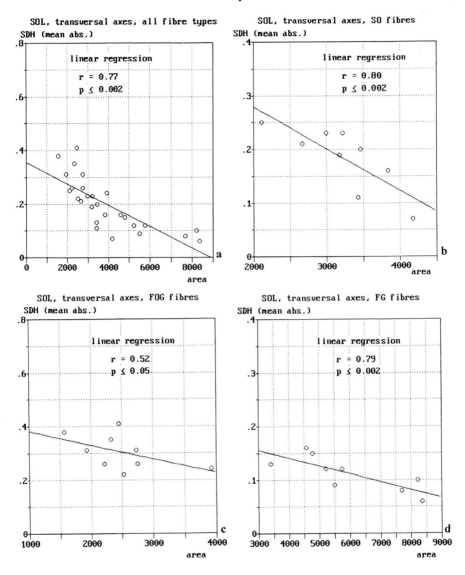

Fig. 14. Correlation between succinate dehydrogenase activity *(SDH)* as mean absorbance *(mean abs.)* of the final reaction product and fibre cross area (area) along the transversal axes in soleus muscle *(SOL)*. Changes in all fibre types together (**a**), and in the separate fibre type *SO* (**b**), *FOG* (**c**), and *FG* (**d**) were correlated. Fibre types: *SO*, slow-oxidative; *FOG*, fast-oxidative glycolytic; *FG*, fast-glycolytic; *r*, correlation coefficient; *p*, probability of error

Fig 15. Fibre type specific changes in fibre cross area (area, given in µm2) along the transversal (superficial-deep) axis near the origin (**a**), in the middle part (**b**) and near the insertion (**c**) of extensor digitorum longus muscle *(EDL)*. Fibre types: *SO*, slow-oxidative; *FOG*, fast-oxidative glycolytic; *FG*, fast-glycolytic. Differences between the muscle regions were analysed with the *t*-test; *$p<0.05$, **$p<0.01$

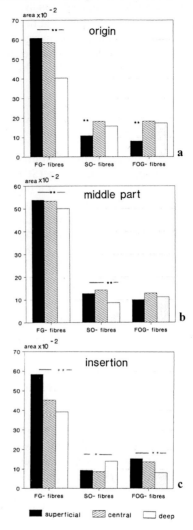

Fibre cross areas along transversal axes of EDL

animal matures, a number of environmental stimuli modulate the fundamental plan. Stimuli include imposed neural activity, as the animal learns increasingly complex skills of locomotion and systemic changes as the mature hormonal milieu unfolds. Therefore, at the time of birth the differentiation into fibre types is not completed, but rather initially determined. The fibre differentiation during the postnatal development as well as the adaptation of fibre type properties to the changing demands for posture and weight bearing increase are worth being investigated and discussed.

The age at which different fibre types can be distinguished histochemically depends on species and muscles. While in man and monkeys the histochemical differentiation takes place before birth, the fibres of hind limbs of newborn rats are histochemically undifferentiated (Schmalbruch 1985b).

In the following we will show that the percentages of slow and fast fibres and of FOG and FG fibres in rat hind limbs change during the postnatal development and ageing. Main attention is focussed on changes in the metabolic profile of the different fibre types during the period of life from the age of 14 to 550 days.

5.2.1
Experimental Basis

The analysis of the following hind limb muscles of the rat is demonstrated: extensor digitorum longus (EDL), soleus (SOL) and gastrocnemius (GAST) muscles. This choice enables the comparison of a fast twitch muscle with predominantly glycolytic metabolism (EDL), a slow twitch muscle with predominantly oxidative metabolism (SOL) and a mixed distal muscle (GAST). The study was performed on 45 rats of both sexes in different age groups given in days (d) with the mean body weight in grams (g) in brackets: 14 d (33 g), 21 d (51 g), 28 d (74 g), 56 d (224 g), 75 d (253 g), 84 d (301 g), 200 d (509 g), 370 d (553 g) and 550 d (625 g). The procedure for muscle analysis was carried out according to the general pattern, see Sect. 5.1.2.

5.2.2
Age-Dependent Changes in Fibre Type Distribution

The investigated hind limb muscles show different fibre type distributions. However, age-dependent changes in fibre type distribution are similar in these muscles (Table 3). The data were graphically demonstrated for EDL as an example (Fig. 16). Only at the age of 14 days did we find two different fibre types, I (slow) and II (fast), after ATPase reaction. At this age fast fibres are a metabolically homogeneous population, all of them showing lower SDH and higher GPDH activity than slow fibres. The fast fibres are still metabolically undifferentiated. The metabolic subtypes FOG and FG come out at the age of 21 days. At the same time also a shift of fibres from slow to fast was observed. After that, during further ageing the slow/fast ratio remains nearly constant. In contrast, the FOG/FG ratio changes continuously from the age of 21 days up to the age of 550 days, i.e. the percentage of FG fibres increases at the expense of FOG fibres.

5.2.3
Age-Dependent Changes of Enzyme Activities in the Different Fibre Types

Not only the population but also the enzyme activities of the fibre types change during postnatal development and ageing. Comparing age-dependent changes of enzyme activities in the different fibre types SO, FOG and FG, differences as well as similarities were detected. Some details from our investigations will be demonstrated in the following. The graphs in Figs. 17–21 arise from data which were obtained from five animals per age.

Table 3. Fibre type distribution in hind limb muscles of rats at different ages

Muscle	Age (Days)	Fibre Types (%)		
		SO I (slow)	FOG II (fast)	FG
EDL	14	7.8±1.0	92.2±1.9 (undifferentiated)	
	21*	5.9±0.1	68.6	25.5±6.7
	28	5.9±0.5	57.8**	36.3±1.2
	75	5.3±0.4	56.6*	38.1±1.7
	84	5.1±1.8	54.4*	40.5±2.7
	200	5.4±0.9	44.6**	50.0±5.7
	370	5.1±0.4	45.9*	48.4±1.5
	550	4.8±0.3	46.2	49.0±3.0
SOL	14	85.4±3.9	14.6±1.8 (undifferentiated)	
	21**	79.1±2.9	18.4	2.5±0.8
	28	80.0±7.1	17.0	3.0±0.6
	75	80.3±7.2	15.6*	4.1±0.6
	84	80.6±5.8	14.9	4.5±0.9
	200	81.8±4.5	13.1	5.1±1.1
	370	81.3±6.2	13.7	5.0±0.9
	550	81.2±5.4	13.8	5.0±0.4
GAST	14	14.6±0.1	85.4±1.6 (undifferentiated)	
	21**	10.2±1.6	77.4	17.4±4.8
	28	12.2±1.8	69.0**	19.0±1.6
	75	12.4±1.7	68.6	19.0±2.0
	84	12.0±0.6	65.2**	22.8±0.8
	200	10.5±0.8	63.8**	25.7±1.4
	370	10.6±1.7	63.2	26.2±1.1
	550	12.3±1.1	61.8	25.9±2.0

Values are means±standard error. Age-dependent differences of fibre type proportion were proved with t-test. Significant differences between the neighbouring age steps were marked by *$p<0.05$, **$p<0.01$. Additionally, all differences in fibre type distribution of a given age and the age of 21 days are significant ($p<0.001$). (The portion of FOG fibres was obtained as difference 100−(FG+SO) without standard error.)

SO fibres of all investigated muscles show age-dependent changes mainly in SDH and ATPase activities, while their GPDH activity remains nearly unchanged up to the age of 200 days (Fig. 17). The maximum of the development-curve is at the age of 21 days. Here maximal SDH and ATPase activities were measured. During the development up to this stage, these enzyme activities increase. After the 21st day, when the animals leave their nest, the demands for posture and mobility of hind limbs change immediately. To be prepared for the higher demands, the oxidative (SDH) activity and the contractile (ATPase) activity of SO fibres have developed their maximal amount. The glycolytic metabolism plays a minor role in SO fibres which are slow twitch fibres with a more supporting function. This may be the reason that no significant changes were found in the glycolytic (GPDH) activity. During the subsequent development and ageing the fibres adapt themselves to the higher demands.

Age-dependent changes in fibre type distribution of EDL

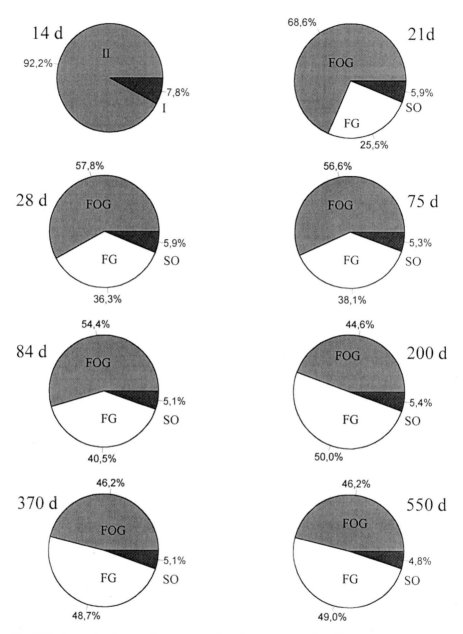

Fig. 16. During ageing from 21 days up to 370 days the percentage of FG fibres increases significantly (shown by t-test, $p<0.01$) at the expense of FOG fibres, the percentage of SO fibres remains nearly constant, demonstrated for extensor digitorum longus muscle *(EDL)*. At the age of 14 days only two different fibre types I (slow) and II (fast) were found. Fibre types: *SO*, slow-oxidative; *FOG*, fast-oxidative glycolytic; *FG*, fast-glycolytic

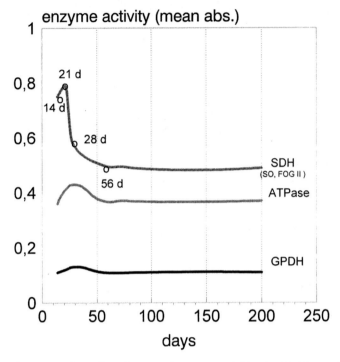

Fig. 17. Age-dependent changes of enzyme activities in SO (slow-oxidative) fibres of extensor digitorum longus muscle *(EDL)*. Note the highest activity level of *SDH* and *ATPase* at the age of 21 days *(d)*. The GPDH activity of SO fibres does not change during ageing. SO and FOG II fibres cannot be differentiated by SDH activity. *SDH*, succinate dehydrogenase; *ATPase*, myofibrillic adenosine triphosphatase ; *GPDH*, glycerol-3-phosphate dehydrogenase; *mean abs.*, mean absorbance

Enzyme activities take their "normal" level which was found in adult stages from 56 days to 200 days. This adult level is lower in EDL than that of the 21st day (Fig. 17). SO fibres from muscles with more supporting function (SOL, GAST) keep the SDH activity level of the 21st day (not shown). At later ages the enzyme activities of SO fibres change once again. At the age of 370 days enzyme activities decrease, and at 550 days all investigated enzyme activities increase significantly, which is demonstrated for SDH activity in Fig. 20. This phenomenon was observed in all investigated muscles and could be interpreted as adaptation of muscle fibres to a larger weight bearing, diminished fitness or changing hormonal milieu during the later age. It is interesting to discuss changes of GPDH and SDH activities in connection, because of the different extent of changes. The GPDH/SDH activity quotient is increased at the age of 370 days and decreased at the age of 550 days compared with the level of the life period from 56 to 200 days. This means that at the age of 370 days, the metabolic profile of SO fibres becomes less oxidative, whereas at the age of 550 days a shift to more oxidative metabolism occurs.

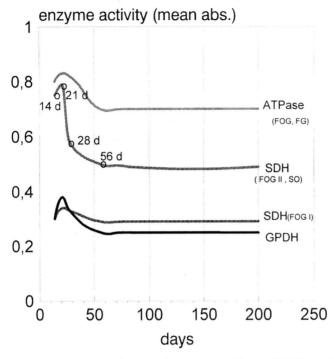

Fig. 18. Age-dependent changes of enzyme activities in *FOG* (fast-oxidative glycolytic) fibres of extensor digitorum longus muscle *(EDL)*. Note the highest level of enzyme activities at the age of 21 days *(d)*. FOG and FG fibres cannot be differentiated by ATPase activity. SO and FOG II fibres cannot be differentiated by SDH activity. *SDH*, succinate dehydrogenase; *ATPase*, myofibrillic adenosine triphosphatase ;*GPDH*, glycerol-3-phosphate dehydrogenase; *mean abs.*, mean absorbance

FOG fibres of all investigated muscles show age-dependent changes in SDH, GPDH and ATPase activities demonstrated for EDL in Fig. 18. (The ATPase activity cannot be measured for FOG fibres separately, only for type II fibres, including both FOG and FG fibres.) Always at the age of 21 days the maximal enzyme activity occurs. The enzyme activities were also compared with that of fast fibres at the age of 14 days. At this age the fast fibres are metabolically undifferentiated and at best include the "precursor fibres" of FOG and FG fibres. FOG fibres can be differentiated at the age of 21 days. They are subdivided according their level of SDH activity in FOG I and FOG II fibres. It should be remembered (see Sect. 4.3) that one part of FOG fibres (FOG I or FOG II, depending on the muscle) and SO fibres have the same SDH activity; they cannot be distinguished from each other by SDH activity alone. Consequently, age-dependent changes in SDH activity are the same in SO fibres and in the overlapping part of FOG fibres. Only the GPDH/SDH-activity quotient makes the differentiation of SO, FOG I and FOG II fibres possible. FOG I and FOG II fibres differ in the level of their SDH activity, but age-dependent changes of SDH activity are similar in both fibre types

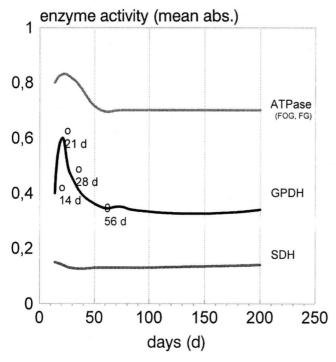

Fig. 18. Age-dependent changes of enzyme activities in *FOG* (fast-oxidative glycolytic) fibres of extensor digitorum longus muscle *(EDL)*. Note the highest level of enzyme activities at the age of 21 days *(d)*. FOG and FG fibres cannot be differentiated by ATPase activity. SO and FOG II fibres cannot be differentiated by SDH activity. *SDH*, succinate dehydrogenase; *ATPase*, myofibrillic adenosine triphosphatase ; *GPDH*, glycerol-3-phosphate dehydrogenase; *mean abs.*, mean absorbance

(Fig. 18). As a consequence of their similar SDH activities, age-dependent changes of SDH activity are similar in FOG and SO fibres. At the age of 21 days FOG fibres of EDL show the maximal SDH activity. After that the SDH activity decreased up to an age of 56 days and remained nearly constant up to 200 days. At later ages the SDH activity of FOG fibres changes. It decreases up to the age of 370 days and then increases up to the age of 550 days (Fig. 20). FOG fibres can work oxidatively as well as glycolytically, they shift between both metabolic pathways to a larger extent than all other fibre types do. Therefore, in contrast to SO fibres, both oxidative and glycolytic enzyme activities play an equal role in the metabolism of FOG fibres. Both enzyme activities change during development and ageing. GPDH and SDH activity of FOG fibres show the same age-dependent course (Fig. 18). At the age of 21 day the highest enzyme activities were found. GPDH activity decreases less drastically than SDH activity. That is the reason why the GPDH/SDH activity quotient increases up to the age of 200 days indicating a shift to more glycolytic metabolism in FOG fibres. One can assume that some of the FOG fibres become more and more glycolytic and transform finally into FG fibres.

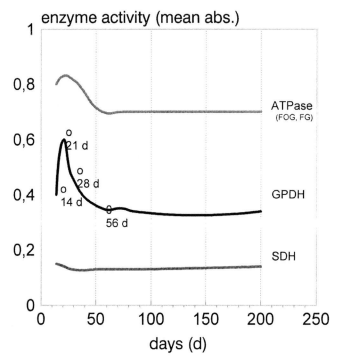

Fig. 19. Age-dependent changes of enzyme activities in *FG* (fast-glycolytic) fibres of extensor digitorum longus muscle *(EDL)*. Note the highest activity level of GPDH and ATPase at the age of 21 days *(d)*. The SDH activity of FG fibres does not change during ageing. FOG and FG fibres cannot be differentiated by ATPase activity. *SDH*, succinate dehydrogenase; *ATPase*, myofibrillic adenosine triphosphatase; *GPDH*, glycerol-3-phosphate dehydrogenase; *mean abs.*, mean absorbance

Actually, as shown in Sect. 5.2.5, the percentage of FG fibres increases during ageing at the expense of FOG fibres.

FG fibres of all investigated muscles show age-dependent changes mainly in GPDH activity and in ATPase activity (analogous to FOG fibres), demonstrated for EDL in Fig. 19. The GPDH activity of FG fibres changes similarly to that of FOG fibres: maximum at 21 days, decrease up to the age of 56 days, constant level up to 550 days. FG fibres have a high glycolytic metabolism, whereas the oxidative ratio of energy production is low. Consequently, the SDH activity or changes in it during ageing plays a minor role in FG fibres. However, at the age of 550 days, the SDH activity of FG fibres increases (Fig. 20), whereas the GPDH activity remains unchanged. That means FG fibres from hind limb muscles of old rats become more oxidative.

Age-dependent metabolic changes of a fibre type do not mean necessarily metabolic changes of the whole muscle. The percentage of the fibre type to the fibre composition and changes in it during ageing is also to be taken into consideration. Considering the EDL for example, we measured biochemically an age-dependent increase in GPDH of

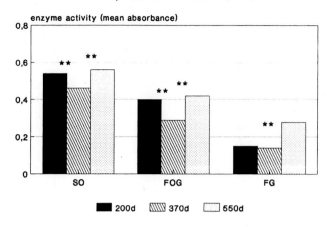

Changes of SDH activity at later ages
in SO, FOG and FG fibres of EDL

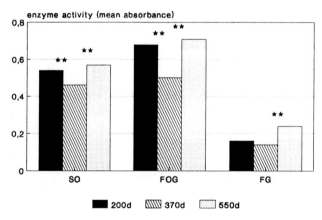

Changes of SDH activity at later ages
in SO, FOG and FG fibres of GAST

Fig. 20. Age-dependent changes of SDH activity in the fibre types of extensor digitorum longus muscle (*EDL*, **a**) and gastrocnemius muscle (*GAST*, **b**). Note the decreased SDH activity of SO and FOG fibres at the age of 370 days *(d)* and the increased SDH activity of FG fibres at the age of 550 days *(d)*. *SDH*, succinate dehydrogenase; fibre types: *SO*, slow-oxidative; *FOG*, fast-oxidative glycolytic; *FG*, fast-glycolytic. *mean abs.*, mean absorbance. ** $p<0.01$

the whole muscle (Fig. 21). On the one hand, the GPDH activity of FG fibres decreases after the 21st day (Fig. 19). On the other hand, the percentage of FG fibres simultaneously increases from 20% to 40% (Table 3). Knowing that the FG fibres are the fibres with the highest glycolytic metabolism, the increasing percentage of FG fibres should be the reason for the age-dependent increase of GPDH activity in the whole muscle.

45

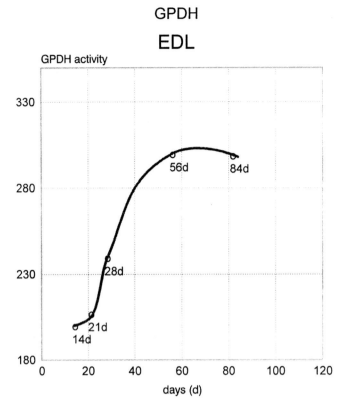

Fig. 21. *GPDH* (glycerol-3-phosphate dehydrogenase) activity (μmol/skg ww) of extensor digitorum longus muscle *(EDL)* during ageing

5.2.4
Conclusions

At the age of 21 days a shift in fibre type population from slow to fast was observed in all investigated rat hind limb muscles. At the same time the metabolic fibre types SO, FOG and FG could be differentiated. Due to transformation of FOG fibres into FG fibres, the FOG/FG ratio decreases during ageing. SO, FOG and FG fibres show changes in their metabolic profile during postnatal development and ageing. That may be interpreted as the adaptation of muscle fibres to changing demands for neural activity, weight bearing, hormonal milieu and other effects. Always the enzyme activity which is characteristic for the fibre type changes: the SDH activity of SO fibres, the GPDH activity of FG fibres, both SDH and GPDH activities in FOG fibres. The ATPase activity changes in both type I (SO) and type II (FG+FOG) fibres.

Age-dependent changes in fibre type related enzyme activities were found to be similar in the investigated muscles. Generally, an increase of enzyme activities was observed up to the 21st day after birth, the time of weaning. After that, the enzyme activities take their adult level which is usually lower than that of the 21st day. In muscles of older rats with an age of 370 and 550 days, further enzyme activity changes of fibre types were measured. Mainly the SDH activity alternates; it decreases at the age of 370 days and increases at the age of 550 days in all fibre types. Age-dependent changes of GPDH/SDH activity quotient in a fibre type give information about changes in the metabolic profile of this fibre type. During the postnatal development the muscle fibres become more glycolytic. During the adult period of 56–200 days the metabolic profile of all fibre types is constant. At the later age of 550 days a shift to more oxidative metabolism was measured mainly in FG fibres.

5.3
Hereditary Myopathy

Myopathic hamsters (BIO8262), often used for investigations of dilatative cardiomyopathy (DCM), have been characterized by Mohr and Lossnitzer 1974. They found morphological alterations at the first week of life in skeletal muscles, and beginning with the 10th week in the myocardium. Clinical symbols like strongly reduced mobility of hind limbs appeared from day 150 to day 200. During this period of ageing, a depressed function of myopathic hamster heart was observed, which was mainly caused by decreased ATPase activities and changes of GPDH and SDH activities (Khuchua et al. 1989; Kjeldsen et al. 1988; Kuo et al. 1992; Malhotra et al. 1985; Norgaard et al. 1987 ; Punkt and Erzen, 2000; Silver and Monteforte 1988). Not only the myocardium, but also skeletal muscles of myopathic hamsters are diseased. Therefore, changes of enzyme activities also occur in skeletal muscles of myopathic hamsters.

We will demonstrate in which way the metabolic profile of different muscle fibre types are effected by myopathy during the period of life from 12–14 to120–190 days. To compare proximal and distal muscles, enzyme activities of fibre types from vastus lateralis and gastrocnemius muscles were measured.

5.3.1
Experimental Basis

The study was performed on 15 myopathic hamsters (BIO 8262) and 15 normal hamsters of both sexes. Myopathic as well as normal hamsters were divided in three groups of age: 12–16, 40–60 and 120–190 days. The hamsters BIO 8262 are charac-terized by hereditary forms of myopathy and cardiomyopathy by autosomal recessive transmission (Mohr and Lossnitzer 1974). Vastus lateralis (VAST) und gastrocnemius (GAST) muscles were prepared. The procedure for muscle analysis was carried out according to the general pattern, see Sect. 5.1.2.

5.3.2
Changes of Enzyme Activities in the Different Fibre Types of Normal and Myopathic Muscles

We will consider changes in the GPDH/SDH activity quotient, because it characterizes changes in the metabolic profile. Figures 22 and 23 show changes of the GPDH/SDH activity quotient in the different fibre types of vastus lateralis and gastrocnemius muscles from normal and myopathic hamsters during ageing. Differences between fibre types on the one hand and between the two muscles on the other hand were measured. The GPDH/SDH activity quotients of FOG and FG fibres of myopathic vastus lateralis muscle are higher than in normal muscle at the first two stages of age and are normalized up to the age of 120–190 days (Fig. 22a,b). In contrast, the SO fibres of myopathic vastus lateralis show differences to that of normal muscles mainly during the later course of ageing (Fig. 22c). From age 42–60 days to age 120–190 days the GPDH/SDH activity quotient significantly decreases in SO fibres of myopathic vastus lateralis muscle. At the age of 120–190 days the GPDH/SDH activity quotient of SO fibres from myopathic vastus lateralis muscle is lower, caused by their higher SDH activity. Comparing these findings with that of myocardium (Punkt et al. 2000), changes in SO fibres of myopathic vastus lateralis are similar to those of myopathic myocardium.

This may be interpreted as the effort of the diseased muscle to compensate the depressed function at the appearance of clinical symptoms.

In contrast to the findings for vastus lateralis muscle, no significant differences between GPDH/SDH activity quotients of fibres from normal and myopathic gastrocnemius muscle were found (Fig. 23).

Generally, after leaving the nest, the metabolic profile of fast fibres (FOG and FG) from normal hind limb muscles is developed more glycolyticly. This enables the generation of force for movement in a faster way (Schmalbruch 1985a). Myopathic changes of metabolic profile were found in the fibres of vastus lateralis muscle as a proximal muscle, but not in the fibres of gastrocnemius muscle as a distal muscle. This suggests that the investigated form of hereditary myopathy effects mainly proximal muscles. It is known that hereditary myopathies "have an unexplained predilection for proximal muscles" (Griggs and Markesbery 1994). Therefore, our findings may be considered reliable. At the first two stages of age, fast fibres of myopathic vastus lateralis are more glycolytic and less oxidative than those of normal hamsters, suggesting less endurance of myopathic than of normal muscles. However, in the later course of age up to 120–190 days, myopathic changes in the metabolic profile were found in slow, but not any longer in fast fibres. Therefore, the clinical symptoms of myopathy, strongly reduced mobility of hind limbs, which were observed from 150th to 200th days, are not due to metabolism of fast twitch fibres but are caused by metabolic changes in SO fibres and other alterations (not shown).

5.3.3
Conclusions

The metabolic profile of a fibre type differs between normal and diseased muscles. Moreover, the fibre type which is mostly effected by myopathy as well as the direction

Fig. 22. Changes in the activity ratio *GPDH* (glycerol-3-phosphate dehydrogenase) and *SDH* (succinate dehydrogenase) of *FOG* (fast-oxidative glycolytic, **a**) *FG* (fast glycolytic, **b**) and *SO* (slow-oxidative, **c**) fibres of vastus lateralis muscle from normal and myopathic hamsters during ageing. Note the differences between fibres from normal and myopathic hamsters. *p<0.05, **p<0.01 and ***p<0.001

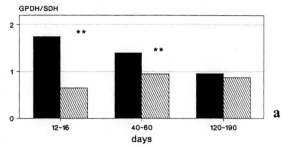

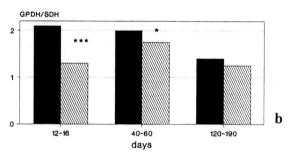

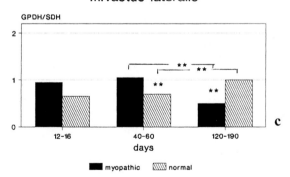

of metabolic shift depend on age, as shown for vastus lateralis muscle from myopathic hamsters. At ages up to 60 days all fibre types of myopathic muscle become more glycolytic. At the age of 120–190 days (onset of clinical symptoms), only the metabolic profile of SO fibres differs between normal and myopathic muscles, it becomes more oxidative. Hereditary myopathy effects enzyme activities in proximal but not in distal muscles.

49

Fig. 23. Changes in the activity ratio *GPDH* (glycerol-3-phosphate dehydrogenase) and *SDH* (succinate dehydrogenase) of *FOG* (fast-oxidative glycolytic, **a**) *FG* (fast glycolytic, **b**) and *SO* (slow-oxidative, **c**) fibres of gastrocnemius muscle from normal and myopathic hamsters during ageing. No significant differences between fibres from normal and myopathic hamsters were found

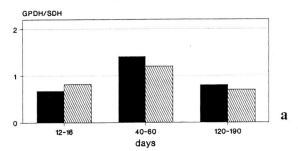

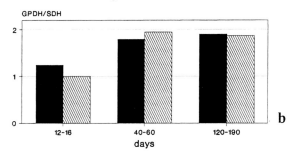

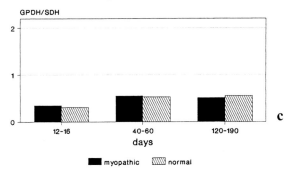

5.4
Experimental Acute Hypoxia

Hypoxia is known to be accompanied by metabolic and functional alterations in heart and skeletal muscles, dependent on the type and duration of hypoxia. Extensive biochemical studies were published in this field, revealing, for instance, an adaptation of the muscle in case of chronic exposure to hypoxia, which results in decreasing oxidative enzyme activities and increasing glycolytic enzyme activities (Taguchi et al. 1985; Howald et al. 1990; Pastoris et al. 1991; Hochachka 1992; Hoppeler and Desplanches 1992; Takahashi et al. 1993). In contrast to chronic hypoxia, acute hypoxia results in an acute response in the muscle, including the increase of oxidative enzyme activities (Veitch et al. 1991; Hoppeler and Desplanches 1992). Additionally, the metabolic and functional reactions caused by hypoxia depend on the metabolic and physiological character of the muscle investigated, as shown, e.g. by Jozsa et al. (1985) for the extensor digitorum longus and soleus muscles of the rat. Not only the hypoxic effect on the whole muscle, but also the hypoxic effect on metabolically and physiologically different muscle fibre types is an interesting problem. We will go through this problem in the next paragraphs. In this context, the fibre type-specific protection of enzyme activities by an antioxidant during hypoxia is particularly interesting.

In the course of this chapter we will answer the following questions:
1. Does acute hypoxia give rise to a metabolic shift within the fibres?
2. In which way are hypoxic effects different in various types of muscles and muscle fibres?
3. Does Ginkgo biloba extract protect the muscle fibres against hypoxia?

5.4.1
Experimental Basis

The investigations were performed on 18-month-old rats; they were divided into three groups. Group 1: six rats without any special treatment served as controls. Group 2: six rats exposed to experimental hypoxia. The animals were placed in a hypoxia chamber consisting of a glass exsiccator combined with an anaesthetic apparatus. An isobaric N_2/O_2-mixture was let in, initially with 20.9 vol % O_2, after 5 min reduced to 5 vol % O_2 for 13 min, and then continuously reduced to 0 vol % during 2 min. After approximately 1 min the respiration ceased. The animals were removed and sacrificed. Group 3: six rats were daily treated with 10 mg per 100 g body weight Ginkgo biloba extract (EGb 761), dissolved in 30 ml water per day for 3 months before exposure to hypoxia in the same manner as in group 2. Animals were sacrificed by a blow on the head and then decapitated. The soleus (SOL) and extensor digitorum longus (EDL) muscles were prepared and frozen in liquid nitrogen. Samples of both muscles of the rats from the three groups were mounted together on one cryostat table and treated according to the general pattern, see Sect. 5.1.2. In this way, tissue sections of rats in the three groups were comparable, resulting in six different samples on one glass slide. Variations caused by differences in section thickness and incubation conditions were avoided, at least for those sections on the same slide. For more details see Punkt et al. 1996.

5.4.2
Hypoxia-Dependent Changes of Enzyme Activities in the Different Fibre Types Without EGb 761 Pretreatment.

The influence of hypoxia on enzyme activities of the fibre types in SOL and EDL is shown in Figs. 24–31. In soleus muscle, only SDH activity is significantly changed by hypoxia in all fibre types. SO and FOG fibres are more affected than FG fibres (Fig. 25). This is also reflected in the decrease of the GPDH/SDH activity ratio mainly in SO and FOG fibres (Fig. 27). In contrast to SDH activity, GPDH activity is not changed (Fig. 26). A significant increase of ATPase and SDH activities is observed in all fibre types of extensor digitorum longus muscle (Figs 28 and 29). The GPDH activity is significantly increased in SO and FOG fibres only (Fig. 30). Similarly to soleus muscle, larger changes in activity were also detected in SO and FOG fibres than in FG fibres in extensor digitorum longus muscle. However, after hypoxia the activity quotient is nearly unchanged in SO and FG fibres and decreased in FOG fibres (Fig. 31).

Generally, whenever a hypoxic effect on muscle fibres is observed, it is a small increase of enzyme activity. This could be explained by the stimulation of enzyme activities in the initial stage of hypoxia, when the ATP content is still sufficient (Freisleben et al. 1991). The investigated enzymes are located at the inner mitochondrial membrane (Stoward et al. 1991) or in the myofilaments (Borgers et al. 1991). They take part in mitochondrial respiration which was found to be higher after short periods of hypoxia or ischemia in rat myocardium (Veitch et al. 1991). This could also sustain the increased enzyme activities in the skeletal muscle after short hypoxia, reflected mainly in SO and FOG fibres. The SDH activity is the most affected parameter. It can be expected that in the later course of hypoxia the SDH activity will decrease (Freisleben et al. 1991). On the basis of in situ analysis of cellular metabolism (van Noorden and Jonges 1995), fibres with low oxygen capacity (FG) will be more vulnerable for low oxygen levels than fibres with high oxidative capacity (SO and FOG), which should enable efficient use of oxygen. In contrast to the soleus muscle, in

Fig. 24. The activity of myofibrilic ATPase (adenosine triphosphatase) in the fibre types of soleus muscles of rats from the different experimental groups. Fibre types: *SO*, slow-oxidative; *FOG*, fast-oxidative glycolytic; *FG*, fast-glycolytic. Differences between the groups are shown with the *t*-test;**$p<0.01$

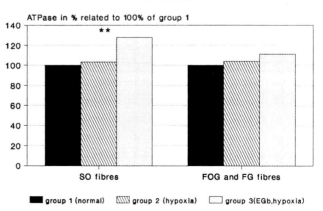

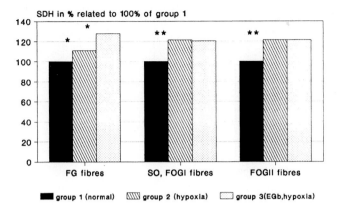

Fig. 25. Changes in activity of SDH (succinate dehydrogenase) in the fibre types of soleus muscles of rats from the different experimental groups. Note the significant increase of SDH activity in all fibre types after hypoxia. Fibre types: *SO*, slow-oxidative; *FOG I*, fast-oxidative glycolytic fibre type with moderate SDH activity; *FOG II*, fast-oxidative glycolytic fibre type with high SDH activity; *FG*, fast-glycolytic. Differences between the groups are shown with the *t*-test; *$p<0.05$, **$p<0.01$

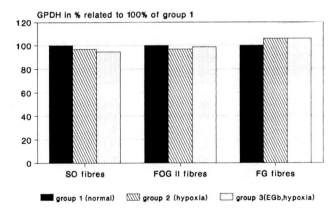

Fig. 26. Unchanged GPDH (glycerol-3-phosphate dehydrogenase) activity in the fibre types of soleus muscles of rats from the different experimental groups. Fibre types: *SO*, slow-oxidative; *FOG*, fast-oxidative glycolytic; *FG*, fast-glycolytic

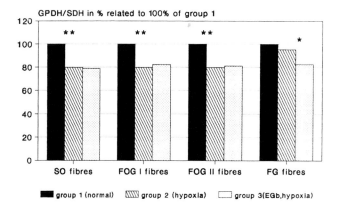

Fig. 27. Changes in the activity ratio *GPDH* (glycerol-3-phosphate dehydrogenase) and *SDH* (succinate dehydrogenase) in the fibre types of soleus muscles of rats from the different experimental groups. Note the decreased GPDH/SDH activity ratio in SO and FOG fibres after hypoxia with and without EGb 761 treatment.. Fibre types: *SO*, slow-oxidative; *FOG I*, fast-oxidative glycolytic fibre type with moderate *SDH* activity; *FOG II*, fast-oxidative glycolytic fibre type with high SDH activity; *FG*, fast-glycolytic. Differences between the groups are shown with the *t*-test; *$p<0.05$, **$p<0.01$

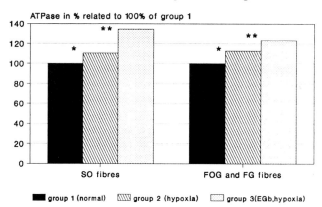

Fig. 28. Increased activity of myofibrillic ATPase (adenosine triphosphatase) in the fibre types of extensor digitorum longus muscles of rats after hypoxia with and without EGb 761 treatment. Fibre types: *SO*, slow-oxidative; *FOG*, fast-oxidative glycolytic; *FG*, fast-glycolytic. Differences between the groups are shown with the *t*-test; *$p<0.05$, **$p<0.01$

SDH
m. extensor digitorum longus

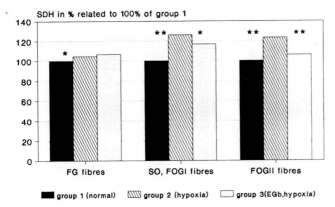

Fig. 29. Changes in activity of *SDH* (succinate dehydrogenase) in the fibre types of extensor digitorum longus muscles of rats from the different experimental groups. Note the significant increase of SDH activity in SO and FOG fibres after hypoxia and the decreasing effect of EGb 761. Fibre types: *SO*, slow-oxidative; *FOG I*, fast-oxidative glycolytic fibre type with moderate SDH activity; *FOG II*, fast-oxidative glycolytic fibre type with high SDH activity; *FG*, fast-glycolytic. Differences between the groups are shown with the *t*-test; *$p<0.05$, **$p<0.01$

GPDH
m. extensor digitorum longus

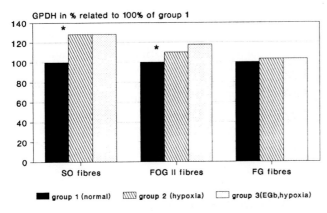

Fig. 30. Changes in activity of *GPDH* (glycerol-3-phosphate dehydrogenase) in the fibre types of extensor digitorum longus muscles of rats from the different experimental groups. Note the significant increase of GPDH activity in SO and FOG fibres after hypoxia. Fibre types: *SO*, slow-oxidative; *FOG*, fast-oxidative glycolytic; *FG*, fast-glycolytic. Differences between the groups are shown with the *t*-test; *$p<0.05$

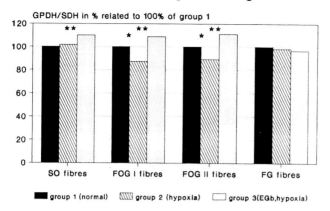

GPDH/SDH activity ratio
m. extensor digitorum longus

GPDH/SDH in % related to 100% of group 1

group 1 (normal)   group 2 (hypoxia)   group 3(EGb,hypoxia)

Fig. 31. Changes in the activity ratio *GPDH* (glycerol-3-phosphate dehydrogenase and *SDH* (succinate dehydrogenase) in the fibre types of extensor digitorum longus muscles of rats from the different experimental groups. Note the decreased GPDH/SDH activity ratio in FOG fibres after hypoxia and the increasing effect of EGb 761. Fibre types: *SO*, slow-oxidative; *FOG I*, fast-oxidative glycolytic fibre type with moderate SDH activity; *FOG II*, fast-oxidative glycolytic fibre type with high SDH activity; *FG*, fast-glycolytic. Differences between the groups are shown with the *t*-test;*$p<0.05$, **$p<0.01$

the extensor digitorum longus muscle not only the SDH activity, but also the GPDH and ATPase activities increase in SO and FOG fibres, suggesting a higher variability of the glycolytic capacity and the contractility in the fibres of a fast twitch muscle with predominantly glycolytic metabolism. Fibre type-specific correlated changes of SDH and GPDH activities are reflected by changes of the GPDH/SDH activity ratio, representing alterations of the metabolic profile in the given fibre type. The decrease of activity ratio seen mainly in the SO and FOG fibres of soleus muscle indicates a shift to higher oxidative ratio of energy production within the oxidative fibres. With respect to the extensor digitorum longus muscle, a small shift towards oxidative metabolism is only observed in FOG fibres. The FOG fibres seem to be the most metabolically adaptable fibre type.

5.4.3
Hypoxia-Dependent Changes of Enzyme Activities in the Different Fibre Types After EGb 761 Pretreatment

Oxygen radicals formed during hypoxia can damage mitochondrial membranes and enzymes and may destroy ATP. Furthermore, intracellular antioxidants were shown to be diminished after 20-min hypoxia in rat hearts (Kirschenbaum and Singal 1992). Consequently, to protect muscle tissue against hypoxia, free radical formation should be reduced and concentrations of antioxidants should be increased. Earlier, we have shown the protective effect of the Ginkgo biloba extract (EGb 761) on hypoxic rat myocardium (Punkt et al. 1995). EGb 761 is a well defined antioxidant (Drieu 1988) with radical scavenging abilities (Pincemail and Deby 1988) and is used in medical therapy (Guillon et al. 1988).

After pretreatment with EGb 761 (group 3) the SDH activity of FG fibres of soleus muscle increases (Fig. 25). Therefore, the activity ratio of FG fibres of soleus muscle becomes lower (Fig. 27). Moreover, an ATPase-increasing effect of EGb 761 is observed in type I (=SO) fibres of soleus as well as of extensor digitorum longus muscle and in type II fibres (including FOG and FG) of the extensor digitorum longus muscle (Figs. 24 and 36). The free radical scavenger EGb 761 may protect ATP which could cause the increase of ATPase activity. Evaluation of serial cross sections revealed that the ATPase activity of FOG fibres is changed more than that of FG fibres. Effects of EGb 761 on SDH activity were mainly found in the SO and FOG fibres of extensor digitorum longus muscle (Fig. 29), whereas the GPDH activity does not change significantly (Fig. 30). The GPDH/SDH activity ratio of SO and FOG fibres of extensor digitorum longus muscle increases after EGb 761 pretreatment (Fig. 31). These findings show that the oxidative fibres are the fibres which are most influenced by EGb 761. A primary effect of EGb 761 on different fibres is observed after a short period of hypoxia: The SO and FOG fibres of extensor digitorum longus muscle show decreased SDH activity in comparison to hypoxic alterations without EGb 761 pretreatment. This is interpreted as a shift to normal metabolism. In contrast to extensor digitorum longus muscle, the SDH activity of oxidative fibres of soleus muscle are not effected by EGb 761 during the primary stage of hypoxia (Fig. 25). We assume that in the soleus muscle, which consists mainly of oxidative fibres, the effect of EGb 761 on the metabolism of these fibres could be observed in later stages of hypoxia.

5.4.4
Conclusions

The metabolic properties of muscle fibres change during acute hypoxia. These changes are not equal in all fibres. Different effects of hypoxia were shown in different fibre types. Answering the questions put at the beginning of Sect. 5.4, we can state:

1. After acute short hypoxia fibre type-specific increase of enzyme activities and a shift to higher oxidative metabolism are observed.
2. The metabolic capacity of oxidative fibre types SO and FOG increases in both soleus and extensor digitorum longus muscles. The metabolic shift to more oxidative metabolism occurs mainly in the fibres of soleus muscle. Hypoxic alterations of enzyme activities are dependent on the metabolic and physiological character of muscle.
3. EGb 761 has a protective effect against hypoxia on the oxidative fibre types SO and FOG.

The increased ATPase activity in oxidative fibres of both muscles and the shift of SDH activity to normal metabolism in the oxidative fibres of extensor digitorum longus muscle may be interpreted as a protective effect of EGb 761 against hypoxia.

5.5
Streptozotocin-Induced Diabetes

Metabolism and function of skeletal muscles are changed during experimentally induced diabetes. There have been reports of altered ratios between glycolytic and oxidative enzyme activities in muscle of rat and men (MacDonald and Swan 1992; Cotter et al. 1993; Eriksson et al. 1994; Kainulainen et al. 1994; Mc Cabe 1994; Simoneau 1998). Differences in diabetic myopathy were found when the fibre composition is different. Kainulainen et al. (1994) found reduced glucose-uptake in muscles containing mainly type I-fibres, but no effect in muscles containing mainly type II-fibres. Atrophy of muscle fibres and a shift in fibre type population caused by streptozotocin (STZ) – as well as alloxan-diabetes – were mainly found in fast muscles (Grossie 1982; Paulus and Grossie 1983; Cotter et al. 1993; Klueber and Feczko 1994). In this way, diabetic alterations were shown for the whole muscle. In the following we will demonstrate the effects of diabetes on defined fibre types.

5.5.1
Experimental Basis

The investigations were performed on 20 male 6-month-old Wistar rats, strain Crl: (Wi) Br, from Charles River GmbH (Sulzfeld, Germany), kept under standardized conditions. The rats were divided into four groups. Group1, control : five rats without any special treatment served as controls. The blood glucose levels of these rats were determined to be 9–16 mmol/l. Group 2, diabetes: Diabetes was induced by intraperitoneal administration of 60 mg/kg body weight STZ (Boehringer, Mannheim, Germany) dissolved in 0.1 M citrate buffer pH 4.5 immediately before use. Five rats at the age of 2 months were subjected to injections of STZ in the morning following a night without food. After 12 days, blood glucose levels were determined to be 27–33 mmol/l. After 4 months of diabetes (blood glucose concentrations >33.3 mmol/l) the rats were sacrificed for analysis. Group 3, control treated with Ginkgo biloba extract(EGb 761, IPSEN, Paris, France): five rats (3 months old) were treated daily with 100 mg/kg body weight Ginkgo biloba extract) during 3 months. Group 4, diabetes treated with EGb 761: five diabetic rats like in group 2 at the age of 3 months, were treated with EGb 761 in the same manner as the rats in group 3. Animals were anaesthetized with ether and then decapitated. Extensor digitorum longus (EDL) and soleus (SOL) muscles were removed, muscle samples from the middle portion between origin and insertion of muscle were prepared, powdered with talcum and frozen in liquid nitrogen. Muscle samples of rats of the four experimental groups were mounted together on a cryostat chuck and treated according to the general pattern, see Sect. 5.1.2. By placing the muscle samples of the four groups on one slide, variations were avoided that are caused by differences in section thickness and incubation conditions.

Distribution, minimal diameters and cross section areas of fibres were measured with the Imaging system KS 100 (Kontron, Eching, Germany). Moreover, myosin heavy chain electrophoresis was performed. For more details in methods, see Punkt et al. 1999.

5.5.2
Effects of Diabetes on Enzyme Activities in the Different Fibre Types

Extensor Digitorum Longus Muscle. GPDH activity increases in FOG and FG fibres by 20%–30% and does not change in SO fibres due to induction of diabetes (Fig. 32a). SDH activity increases by nearly 30% in SO and FG fibres and is unchanged in FOGII fibres (Fig. 32b). Consequently, the GPDH/SDH ratio increases in FOG, decreases in SO and remains unchanged in FG fibres (Fig. 32c). **Soleus muscle.** Enzyme activities are increased in FOG and FG fibres similar to that of EDL. Additionally, SDH activity is also higher in FOGII fibres by approximately 10% (Fig. 33a,b). SO fibres of SOL are affected more by diabetes than those in EDL. SO fibres of SOL show an increase in GPDH activity by 13% and in SDH activity by 41%. Changes in GPDH/SDH ratios are similar to those of EDL, increased in FOG and decreased in SO fibres. In SO fibres the ratio is decreased more than in EDL (Fig. 33c).

Increased enzyme activities in the different fibre types are interpreted as metabolic compensation of diabetic muscles for reduced performance. This agrees with our findings for the diabetic heart, showing increased enzyme activities (Punkt et al. 1997). It may be assumed that in both skeletal muscle and heart muscle oxidative stress and oxygen consumption are increased (Rösen et al. 1986). Uptake of glucose into myocytes is reduced, but the utilization of available glucose is increased (Rösen et al. 1992). The more effective utilization may be realized by the higher enzyme activities that were measured in the present study. Changes in GPDH/SDH activity ratio of a given fibre type characterize changes in the metabolic profile of this fibre type. In this way, FOG-fibres of diabetic EDL and SOL muscles show a shift to more glycolytic metabolism, in contrast to SO fibres which show a shift to more oxidative metabolism. The metabolic profile of FG fibres remains unchanged in both muscles. Evidently, diabetes causes a metabolic shift in oxidative muscle fibres only.

Besides similarities, also variations in the two muscle types investigated were found. The increase in enzyme activities in oxidative fibres is higher in diabetic SOL than in diabetic EDL, suggesting stronger effects of diabetes or a more effective response to diabetes in muscles with a predominant oxidative metabolism. Furthermore, the stronger effect of diabetes in SOL than in EDL support the findings of Kainulainen et al. (1994) who showed reduced glucose-uptake to occur mainly in slow muscles, such as SOL.

5.5.3
Effects of EGb 761 on Enzyme Activities of Diabetic Muscles

Extensor Digitorum Longus Muscle. The strongest effect of EGb 761 was found to be exerted on GPDH activity in SO fibres of diabetic muscle. The GPDH activity was increased by 50%, whereas GPDH activity was not significantly changed in untreated diabetic rats (Fig. 32a). Additionally, a further increase in GPDH activity was found in FOG fibres (14%) and FG fibres (8%) of diabetic muscle after EGb 761 treatment in comparison to the elevated GPDH activity in untreated diabetic muscle. The increase of SDH activity was similar in untreated and treated diabetic rats (Fig. 32b).

Extensor digitorum longus muscle

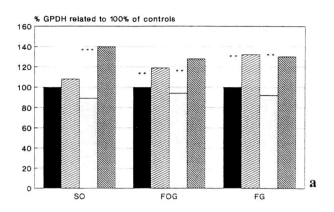

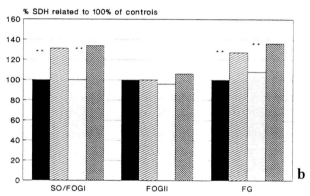

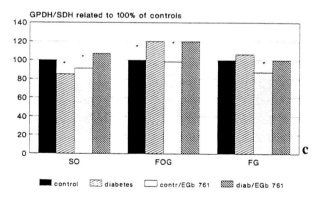

control   diabetes   contr/EGb 761   diab/EGb 761

Fig. 32. Changes in activity of GPDH (glycerol-3-phosphate dehydrogenase, **a**), and SDH (succinate dehydrogenase, **b**) and in the GPDH/SDH activity ratio (**c**) in the fibre types of extensor digitorum longus muscle of rats from different experimental groups. Fibre types: *SO*, slow-oxidative; *FOG I*, fast-oxidative glycolytic fibre type with moderate SDH activity; *FOG II*, fast-oxidative glycolytic fibre type with high SDH activity; *FG*, fast-glycolytic. FOGI and SO fibres show similar SDH activity and cannot be differentiated by SDH activity only. Differences between the control and diabetic rats and between the EGb 761 treated control *(contr/EGb 761)* and diabetic *(diab/EGb 761)* rats are shown with the *t*-test; $^*p<0.05$, $^{**}p<0.01$, $^{***}p<0.001$

Soleus muscle

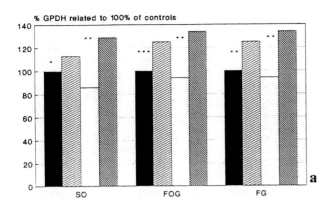

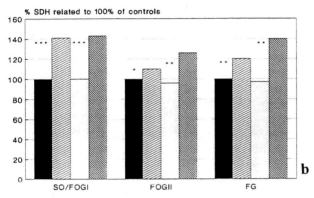

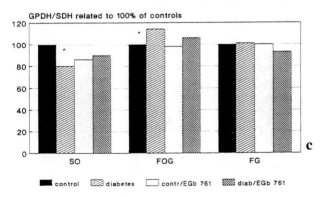

Fig. 33. Changes in activity of GPDH (glycerol-3-phosphate dehydrogenase, **a**), and SDH (succinate dehydrogenase, **b**), and in the GPDH/SDH activity ratio, (**c**), in the fibre types of soleus muscle of rats from different experimental groups. Fibre types: *SO*, slow-oxidative; *FOG I*, fast-oxidative glycolytic fibre type with moderate SDH activity; *FOG II*, fast-oxidative glycolytic fibre type with high SDH activity; *FG*, fast-glycolytic. FOGI and SO fibres show similar SDH activity and cannot be differentiated by SDH activity only. Differences between the control and diabetic rats and between the EGb 761 treated control *(contr/EGb 761)* and diabetic *(diab/EGb 761)* rats are shown with the *t*-test; $*p<0.05$, $**p<0.01$, $***p<0.001$

Therefore, the GPDH/SDH activity ratio increased in all fibre types of diabetic muscle after EGb 761 treatment (Fig. 32c).

Soleus Muscle. The strongest effects of EGb 761 were found in SO fibres where GPDH activity increased by 30% (Fig. 33a) and in FOGII fibres where SDH activity increased by 20% (Fig. 33b), in comparison to the increase of enzyme activities in untreated diabetic rats. Consequently, the changes in GPDH/SDH activity ratio which were observed in SO and FOG fibres of untreated diabetic rats diminished after EGb 761 treatment (Fig. 33c). GPDH and SDH activities were more strongly increased in FG fibres of EGb 761-treated than in untreated diabetic rats (Fig. 33a,b). Both enzyme activities increased in a similar way. Therefore, GPDH/SDH activity ratios remained unchanged.

Free oxygen radicals formed in diabetic muscles can damage mitochondrial membranes and enzymes located there (Gerber and Siems 1987). EGb 761 is a potent scavenger of oxygen radicals (Pincemail et al. 1988) and should protect mitochondrial enzyme activities in the fibres of diabetic muscles. After EGb 761 treatment, the increase in enzyme activities was stronger in all fibres of both muscles. This may be interpreted as an increase in metabolic reserve and therefore as a protective effect of EGb 761 on diabetic muscles. Moreover, in contrast to untreated diabetic rats, a shift to more glycolytic metabolism was found in all fibre types in EDL of treated diabetic rats. In SOL the metabolic character of fibres remained unchanged in comparison to controls. The strongest effect of EGb 761 on diabetic muscles was the increased GPDH activity in SO fibres of both EDL and SOL and the increased SDH activity in FOG fibres of SOL. Evidently, EGb 761 affects mainly oxidative fibres, as already shown in a recent study (Punkt et al. 1996). So, the metabolic profile of oxidative muscle fibres was shown to be more vulnerable to diabetes than that of glycolytic fibres on the one hand, but EGb 761-protection was most effective in these fibres on the other.

5.5.4
Changes in Fibre Type Distribution, Fibre Cross Areas and MHC-Isoforms of Diabetic Muscles

The fibre type distributions of EDL and SOL of rats from different groups are demonstrated in Fig. 34. In both muscle types, the percentage of SO fibres remains unchanged in all groups. EDL as a fast twitch muscle consists of 5%–8% SO fibres, while SOL as a slow twitch muscle contains 80% SO fibres. In contrast to SO fibres, populations of FOG and FG fibres change under the various conditions. Muscles of nearly all experimental groups contain fewer FOG fibres and more FG fibres in comparison with controls.

When comparing cross sections of muscles of normal and diabetic rats, considerable differences become apparent (Fig. 35a,b). In diabetic muscle, morphology is altered and all fibre cross areas are smaller than in normal muscle. Measurements of the minimal fibre diameters reveal a decrease of 29% (EDL) and of 33% (SOL) in SO fibres and a decrease of 40% (EDL) and of 20% (SOL) in FG fibres of diabetic muscles. After EGb 761 treatment similar effects are observed (Fig. 36a,b). Moreover, the percentage area of a fibre type of the whole muscle cross section area changes in diabetic muscles. A diminished percentage area of SO fibres was found in diabetic EDL and SOL

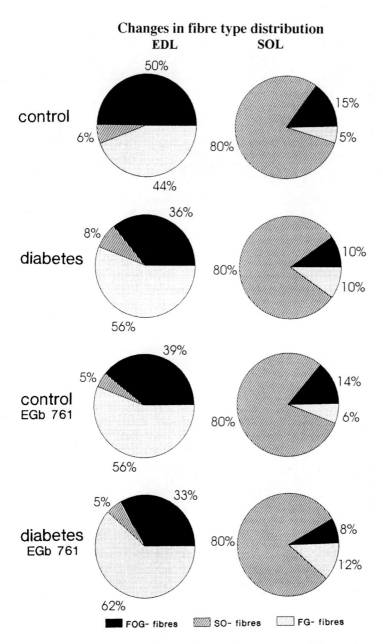

Fig. 34. Changes in the fibre type distribution of *EDL* (extensor digitorum longus) and *SOL* (soleus muscle) of rats from different experimental groups. Fibre types: *SO*, slow-oxidative; *FOG*, fast-oxidative glycolytic; *FG*, fast-glycolytic. Note the changes in percentages of FOG and FG fibres in different groups. Differences in FG population between the control and diabetic rats and between the EGb 761 treated control *(control EGb 761)* and diabetic *(diabetes EGb 761)* rats are shown to be significant ($p < 0.05$)

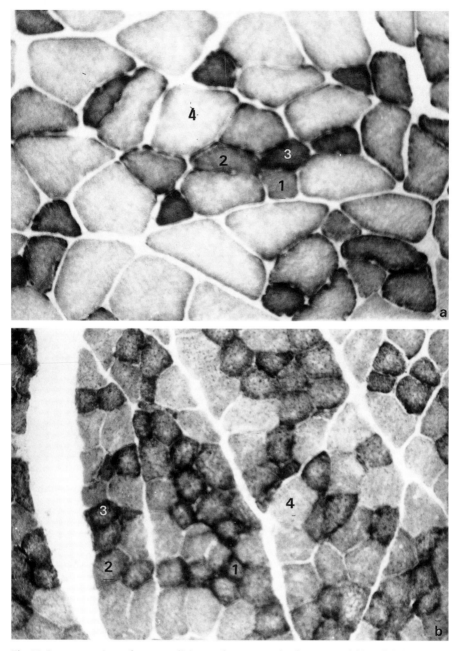

Fig. 35. Cryostat sections of extensor digitorum longus muscles from normal (**a**) and diabetic (**b**) rats reacted for SDH. ×175. Marked fibres: 1=SO, 2=FOG I, 3=FOG II, 4=FG

Minimal fibre diameter

extensor digitorum longus muscle

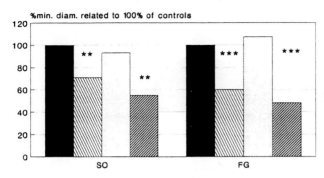

soleus muscle

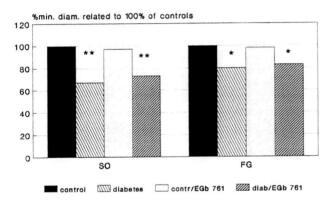

Fig. 36. Changes in minimal fibre diameter *(min. diam.)* of *SO* (slow-oxidative) and *FG* (fast-glycolytic) fibre types of extensor digitorum longus muscle (**a**) and soleus muscle (**b**) of rats from different experimental groups. Note the decreased minimal fibre diameter in diabetic muscles. Differences between the control and diabetic rats and between the EGb 761 treated control *(contr/EGb 761)* and diabetic *(diab/EGb 761)* rats are shown with the *t*-test; *$p<0.05$, **$p<0.01$, ***$p<0.001$

muscles (Fig. 37). Consequently, the percentage area of fast fibres is increased, caused by FOG fibres.

Fig. 38 shows the electrophoretic separation of the different MHC isoforms in control and diabetic EDL and SOL muscles. The diaphragm was used as a reference to identify MHC isoforms. The fastest gel migration is exhibited by the slow MHC I and the lowest by fast MHCs IIA+IID, which form an incompletely separated band. In EDL fast MHC IIB isoforms are located between MHC I and MHC IIA+D. The proportions of MHC isoforms are given in Table 32. Both diabetic muscles show reduced slow and

Cross section area of a fibre type as percentage of the whole muscle cross section area

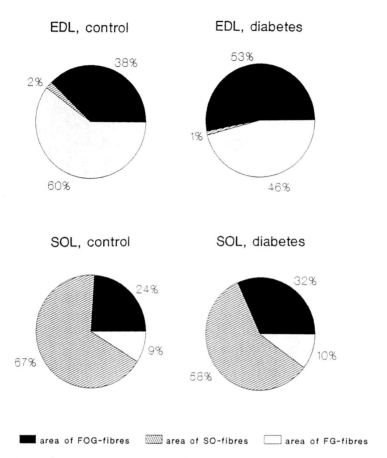

Fig. 37. Changes in percentage area of a fibre type related to the whole muscle cross section in diabetic muscles in comparison with control muscles. *EDL*, extensor digitorum longus muscle; *SOL*, soleus muscle. Fibre types: *SO*, slow-oxidative; *FOG*, fast-oxidative glycolytic; *FG*, fast-glycolytic. Differences between area percentage of SO fibres of control and diabetic muscles and between area percentage of FG fibres of control and diabetic EDL are shown to be significant ($p<0.01$)

increased fast MHC isoforms (mainly IIB in EDL, IIA in SOL) as compared with controls.

The interpretation of the results may be as follows: The proportion between the number of slow and fast fibres remained unchanged in diabetic muscles. Therefore, the reduction in slow MHC I isoforms in both muscles is most likely not due to a reduced number of slow fibres but rather is a result of the reduced percentage areas of SO fibres. The reduced percentage area of a fibre type indicates a relatively reduced

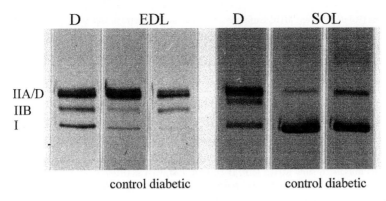

Fig. 38. Silver-stained SDS-Page of MHC isoforms in control and diabetic extensor digitorum longus *(EDL)* and soleus *(SOL)* muscles. Diaphragma *(D)* was used as reference muscle to identify the MHC isoforms: I, slow myosin, II B, II A/D fast myosins

mass of this fibre type. SO fibres are identical with type I-fibres which are characterized by the expression of the slow MHC I isoform. Therefore, reduced MHC I isoforms in diabetic muscle could be caused by atrophy of SO fibres. It may be assumed that the changes in fast MHC isoforms of diabetic muscles are also based on altered mass percentages of corresponding fibre types (IIA, IIB, IID). These fast fibre types which are related to myosin isoforms are not completely identical with the metabolically different fast fibre types FOG and FG that were analysed in the present study. We found changes in percentages of both area and number of FOG and FG fibres in diabetic muscles. Diabetes as well as treatment with EGb 761 cause changes in the FOG/FG ratio. The number of FOG fibres decreases as much as the FG fibres increases. We assume that the shift to glycolytic metabolism as shown above causes the shift from FOG fibre to FG fibre. Evidently, fast fibres adapt themselves to changed conditions by varying their metabolic profile. As already shown in earlier studies about ageing and hypoxia (Punkt et al. 1993 1996), FOG fibres are the most adaptable fibre type. Furthermore, changes in FOG/FG ratio suggest that the classification of FOG and FG fibres is dynamic. Pathophysiological mechanisms for FOG α FG transformation can be the following: FOG and FG fibres belong to different motor units, FOG to FR (fast twitch fatigue resistant) and FG to FF (fast twitch fatiguable) motor units (Burke et al. 1983). On the one hand, we observed that diabetic rats moved less well than healthy rats. They were apathetic. Therefore, fewer motor units may be recruited in diabetic EDL and SOL muscles to realize the diminished motor activity. Fibres of a seldom recruited motor unit, e.g. FOG fibres, show a shift to more glycolytic metabolism, and motor unit changes into a FF motor unit containing FG fibres.

On the other hand, changes of fast fibre populations as well as MHC isoforms in diabetic muscle can also be a response to diabetic neuropathy. It is known that diabetes induces neuropathy which appears in several forms, e.g. a muscle atrophic form (Dyck 1994). Peripheral nerves are damaged in diabetes. This could be the reason for morphological changes such as fibre atrophy in diabetic muscles as observed in the present study (see also Klueber and Fecko 1994). When a nerve of a motor unit is completely destroyed, the motor unit may be reinnervated which can

Table 4. Proportions (%) of different MHC isoforms of control and diabetic muscles

	EDL			SOL	
	I	II B	II A/D	I	II A
Control	3.0±0.4	2.7±0.5	94.3±1.5	89.7±3.1	10.3±1.0
Diabetes	Traces	17.8±1.0	82.2±2.2	79.4±3.2	20.6±1.7

Values are means±SD. Abbreviations: MHC, myosin heavy chain; EDL, extensor digitorum longus muscle; SOL, soleus muscle; I, slow MHC; II B, II D, II A, fast MHCs. Differences between MHC isoforms from control and diabetic muscles are shown to be significant ($p<0.001$).

also lead to a transformation of motor unit and their muscle fibre type, respectively, e.g. from FOG to FG.

We found fibre atrophy in both fast and slow muscles of diabetic rats. However, in diabetic EDL (fast muscle) fast-glycolytic fibres are more atrophied than slow fibres, in contrast to diabetic SOL (slow muscle) where the slow fibres show stronger atrophy. This suggests that muscle fibres of the dominant fibre type of diabetic muscle are most effected in response to diabetic neuropathy. In this context, changes in innervating α-motoneurons should be investigated in further studies.

5.5.5
Conclusions

Streptozotocin-induced diabetes effects the properties of muscle fibre types. Changes in fibre type-related enzyme activities, fibre type distribution, fibre cross areas and myosin isoforms were found. In muscles of diabetic rats, a metabolic shift was measured, mainly in fibres with oxidative metabolism. FOG fibres show a shift to more glycolytic metabolism and about a third of them transform into FG fibres. SO fibres become more oxidative. Fibre atrophy occurs in diabetic muscles dependent on fibre type and muscle. Different fibre types atrophy to a different degree. Therefore, a decreased area percentage of slow fibres and an increased area percentage of fast fibres of the whole muscle cross section in both EDL and SOL muscles are found. This is supported by reduced slow and increased fast myosin heavy chain isoforms. These alterations of diabetic muscle fibres could be due to less motion of diabetic rats and diabetic neuropathy. After treatment with Ginkgo biloba extract, enzyme activities are increased mainly in oxidative fibres of diabetic muscles, which is interpreted as a protective effect. Generally, the soleus muscle with predominant oxidative metabolism is more vulnerable to diabetic alterations and Ginkgo biloba extract treatment than the extensor digitorum longus muscle with predominant glycolytic metabolism.

6 Fibre Type Transformation

Fibre type transformations are the consequence of adaptation of fibres to changed conditions. Muscle fibres are able to modify their molecular composition and their properties by altered gene expression according to the functional demands. The transformation mainly occurs in MHC isoforms. They are adaptive to the physiological changes and can transform in each other. Therefore, the MHC isoform profile is a good marker of muscle fibre type transformation. MHC isoforms expressed in a single fibre can be determined by immunohistochemical techniques. It should be noted that transformation occurs not only in MHC proteins but also in other proteins such as troponin and myosin light chains (Schiaffino and Reggiani 1996).

The regulatory mechanism of the gene expression involved in fibre type transformation is not yet clear. It was shown (Staron and Pette 1987) that the distribution of the MHC isoform along the fibre length is nonuniform even in the skeletal muscle under normal conditions. This fact could explain the changes in enzyme activities along the fibre length that we discussed in Sect. 5.1. The nonuniformity of MHC isoform distribution increases under muscle fibre type transformation (Staron and Pette 1987; Gauthier 1990). The distribution of MHC mRNA along the fibre lenght is also reported to be nonuniform (Peuker and Pette 1997). Under normal conditions there is a correspondence of MHC protein and mRNA (Denardi et al. 1993; Peuker and Pette 1997). However, the content of MHC protein and the expression of mRNA are mismatched in transforming fibres (Bodine and Pierotti 1996; Andersen and Schiaffino 1997). The genetic regulation in the multinuclear skeletal muscle fibres is also a problem, whether a single nucleus produces mRNA of a single isoform or multiple isoforms. Hori et al. (1998) formulated the following questions that are unanswered up until now: "Does the MHC protein produced remain within the nuclear domain of the nucleus which has expressed its mRNA? Does the transformation occur simultaneously along the longitudinal axis of a muscle fiber? If not, where does it start, and how is it coordinated along the fiber's full length?"

Recently, investigations were published (Meißner et al. 2000) to study the signalling pathway underlaying fibre type transitions in primary muscle cell cultures at the mRNA level. The adult fast character of myotubes was demonstrated. A fast-to-slow transition was shown in terms of MHC isoform gene expression after treatment with a Ca^{2+} ionophore. Moreover, up-regulation of citrate synthase mRNA and down-regulation of glyceraldehyde 3-phosphate dehydrogenase mRNA were observed. This points to changes in energy metabolism that are also characteristic of a fast-to-slow transition. Furthermore, the effect of Ca^{2+} ionophore on MHC gene expression proved to be reversible. This study is a contribution to understanding the primary signals that mediate the fibre type transformations. However, in conclusion it was stated that a consistent hypothesis for a signalling pathway underlying fibre type transition has not emerged from this study. The primary skeletal muscle culture offers the possibility to study signalling pathways in vitro and may help to investigate the complete signalling pathways underlying fibre type transformations in vivo.

Several authors (e.g. Pette 1984 1990b; Vrbova et al. 1990; Aigner and Pette 1992; Czesla et al. 1997; Pette 1998; Windisch et al. 1998) investigated the effect of training and chronic electrical stimulation on the contractile apparatus of rat or rabbit skeletal muscles. They found fast-to-slow transitions in the myofibrillar protein isoforms and fibre types. After neuromuscular activity increased, fibre type tranformation from fast to slow respectively from type II to type I was observed, whereas decreased neuromuscular activity induced transformations in the opposite direction (Pette 1984; Pette and Staron 1997). Fibre type transformations from slow to fast were also observed after spinal cord injury in the vastus lateralis muscle of humans (Burnham et al.; 1997). Such fast-to-slow as well as slow-to-fast transitions do not proceed in abrupt jumps from one extreme to the other, but occur in a gradual and orderly sequential manner. Therefore, at all times there also exist hybrid forms of fibres which show co-expression of myosin isoforms. In comparison to that, metabolic transitions underlying the FOG to FG transformation happen earlier and faster (Hoppeler 1990). This means that the myosin isoform pattern of a "new" FG fibre does not necessarily need to be definite but can change in the course of time.

Fibre type transformations were observed in rat hind limb muscles during development and ageing. The fibres of the hind limb muscles of newborn rats show no differences in myofibrillic ATPase activity. The contractile properties of muscle fibres of rats differentiate during the postnatal development and change depending on functional load. Especially at the time of the 21st postnatal day, when the rats leave the nest, the functional load of the hind limbs changes. Schmalbruch (1985b) found that during the period of life from 6 to 140 days the percentage of the fibre types I, IIA, and IIB changes in rat hind limb muscles. The trend differs in slow and fast twitch muscles. For example, in the soleus rat muscle as a slow twitch muscle, fibre type transformations of IIA to I was found during the postnatal development. That can be explained by an increased muscle activity after leaving the nest. When the muscle activity is being reduced by hind limb suspension during postnatal development, the IIA to I transformation is inhibited (Asmussen and Soukup 1991). Moreover, in the case of adult soleus muscle, hind limb suspension transforms the soleus muscle fibre type composition from predominantly type I to approximately equal proportions of type II and type I fibres (Anderson et al. 1999). These results confirm the findings of Pette (1984) that decreased muscle activity favours a slow to fast transition. In the opposite, increased activity induces, in an ordered sequence of events, fast to slow conversions. Increased activity by endurance training leads to changes similar to chronic stimulations (Pette 1998).

In contrast to the IIA to I transformation in the developing slow twitch soleus muscle, a fibre type transformation from I to II was observed in the fast twitch rat extensor digitorum longus muscle and gastrocnemius muscle up to an age of 21 days when the rats leaving the nest (Sect. 5.2 in this book). Schmalbruch (1985b) found IIA to IIB transformations in the fast twitch plantaris muscle of rat up to an age of 140 days. This result is similar to our findings in extensor digitorum longus muscle (see Sect. 5.2) showing FOG to FG transformations during ageing after leaving the nest. The metabolic transformation from FOG to FG becomes definite earlier than the transformation from IIA to IIB which is based on different myosin isoforms.

Fibre type transformation from FOG to FG can be induced either by increased or by decreased motoric activity. On the one hand, fast muscle fibres with high glycolytic metabolism, FG fibres, are necessary to develop fast force for start and spurt. This means that when, after leaving the nest, the rats develop higher motoric activity then more FG fibres are produced by transforming FOG fibres into FG fibres. In the later course of ageing, the motoric activity of rats decreases. Now another effect induces the FOG to FG transformation. For decreased motoric activity less fast motor units of hind limb are necessarily recruited. The muscle fibres of a seldom recruited motor unit become more glycolytic. Consequently, a FR motor unit with FOG fibres can convert into a FF motor unit with FG fibres.

Fibre transformations also occur under pathological conditions when muscle activity and muscle metabolism are altered. One example is the FOG to FG transitions in skeletal muscles of diabetic rats which is presented in Sect. 5.5.

Hormones can induce fibre type transformations. It was shown that growth hormone induces a fast-to-slow transformation in rat soleus muscles (Daugaard et al. 1998). Thyroid hormone inhibits slow MHC isoform expression in rat soleus muscle and favours the appearance of fast MHC isoforms (Soukup and Jirmanova 2000). Hyperthyroidism causes the slow-to-fast transformation, and hypothyroidism causes the slow to fast transformation (Gosselin et al. 1996; Narusawa et al. 1987). Furthermore, it was concluded from muscle grafting that thyroid hormone is necessary for the transformation of the regenerating slow muscle into a fast muscle. That means the transformation of slow fibres into fast fibres. It seems that the thyroid hormone exerts a direct effect on muscles. Different skeletal muscles contain different numbers of thyroid receptors, which result in a different sensitivity of individual muscles to thyroid hormone. Both increased and decreased levels of thyroid hormone lead to changes in muscle fibre compositions (for references see Soukup and Jirmanova 2000).

6.1
Conclusions

Fibre type transformations in skeletal muscles occur under changing conditions. They are induced by several processes such as development, changes in functional demands, changes in neuromuscular activity, training and by hormones. Thus, fibre type transformation is a common phenomenon in skeletal muscles during a lifetime. The signalling pathway underlying fibre type transformation is not fully understood up to now.

Transitions between the physiological slow and fast fibre types were observed in both directions. Generally, increased muscle activity leads to fast-to-slow transitions, and decreased muscle activity induces slow-to-fast-transitions. Such transformations are based on the conversion of myosin isoforms by altered gene expression.

Fibre type transformations happen frequently between histochemical fibre types IIA and IIB and between metabolical fibre types FOG and FG. All of these fibre types belong to fast fibres. One can conclude, that histochemical fibre types represent "metastable entities" within a dynamic equilibrium (Pette 1984), and that fast fibres are a metabolic continuum, where FOG and FG fibres represent the dynamic metabolic situation.

Fibre type transformations reflect the dynamic nature of muscle fibres.

7 FOG Fibres – The Most Adaptable Muscle Fibres

In Chap. 5 we investigated the metabolic changes in muscle fibres under changing conditions. Summarizing the effects of all experiments, one can state that metabolic changes occur predominantly in FOG fibres: After acute hypoxia an increased oxidative activity of FOG fibres was measured, caused by the increased mitochondrial respiration. Diabetes induced a shift to glycolysis in FOG fibres. This was interpreted as a reaction to oxidative stress, changed motoric activity or neuropathy. In diabetic muscles the percentage of FOG fibres decreased to the same amount to which the percentage of FG fibres increased. The same effect was observed in several muscles after treatment of rats with Ginkgo biloba extract and during ageing. We can conclude that FOG fibres change their metabolism as long as they convert into FG fibres. In this way FOG fibres, FOG I as well as FOG II fibres are the most adaptable fibres. That may be caused by their metabolic equipment. FOG fibres are metabolically variable; they have high oxidative as well as glycolytic capacity. This is in contrast to SO and FG fibres where the oxidative and glycolytic capacities are reciprocal. SO and FG fibres prefer one pathway (oxidative or glycolytic, respectively) of energy metabolism; FOG fibres can use both pathways equally. Although each muscle fibre can, to a certain extent, shift between both metabolic pathways, the metabolic shift is possible in FOG rather than in the other fibres. It was shown that under changing experimental conditions, the metabolic shift can finally result in the transition of FOG fibres to FG fibres.

8 Are There Adaptation Processes in α-Motoneurons as in Muscle Fibres?

Adaptation processes may be reflected not only in the muscle fibres, but also in the motoneuron of the motoric unit. Signs of load-specific adaptation in motoneurons were found (Pieper 1984). These are for example activation of motoric cells by increased oxidative and glycolytic metabolism and activation of protein synthesis, increase of nerve conduction velocity and increase of the motoric endplate on the adapted fibre. Almeida et al. (1996) correlated neurophysiological, mechanical and histochemical parameters to demonstrate muscle adaptation with the training of rats. They revealed that neuronal and muscle components of motor units are both affected by training. Recent papers focussed on the problem of whether the α-motoneuron adapts its size and metabolism according to adaption processes of the muscle fibres which are innervated by it. Seburn et al. (1994) measured the SDH activity in rat hind limb muscle fibres and innervating motoneurons under conditions of increased activity over a wide range of activity levels. They found changed SDH activity of type I fibres and unchanged SDH activity of corresponding motoneurons, suggesting that the oxidative activity of motoneurons does not change despite clear adaptations in the muscle fibres they innervate. Nakano et al. (1997) observed similar effects. They measured unchanged SDH activity in motoneurons innervating rat soleus and extensor digitorum longus muscles after chronic activity. However, the size of motoneurons increased for the soleus muscle of trained rats, not for the extensor digitorum longus muscle. It was concluded that chronic activity has a stronger impact of motoneurons innervating slow twitch rather than fast twitch muscles. Ishihara et al. (1995) showed the interdependence in the oxidative capacity between a motoneuron and its target muscle fibres of rat tibialis anterior muscle. Furthermore, soma size of a motoneuron as well as muscle fibre cross section area correlate negatively with SDH-activity in several rat muscles (Ishihara et al 1995; Sect. 5.1, this volume). However, in response to increased neuromuscular activity, muscle fibres and their innervating motoneuron react differently: while the muscle fibres change their size and SDH activity, the motoneurons remain unchanged, measured in fast twitch rat plantaris muscle (Roy et al. 1999).

Summarizing, one can conclude that α-motoneurons, unlike their associated muscle fibres, are stable over a wide range of levels of chronic neuromuscular activity. Adaptation processes of muscle after altered muscle activity may proceed rather in muscle fibres than in the innervating motoneuron.

9 The Influence of Muscle Type on the Properties of a Fibre Type

The investigation of muscles which differ in their physiological and metabolic characteristic has shown: Not only different fibre types, but also different muscle types adapt themselves differently to changed conditions. That means that the properties of a certain fibre type and its adaptability to changed conditions depend on the type of muscle from which the fibres arise. The effects of changed conditions to fibres from different muscles are presented in Chap. 5. Considering the influence of the muscle type on the properties of fibre types, the following effects may be pointed out:

SO and FOG fibres of an oxidative muscle have twice higher cross section areas than SO and FOG fibres of a glycolytic muscle. This may be a means, additionally to the fibre population, to realize the higher oxidative activity of the oxidative muscle than that of the glycolytic muscle. The cross section areas of FG fibres are not much larger in glycolytic than in oxidative muscles.

The adaptation of muscle fibres to varied regional functional demands depends on the physiological and metabolic properties of the muscle to which they belong. The fibres of EDL, which is a fast-force muscle with predominant glycolytic metabolism, develop their highest glycolytic activity near the insertion. This may be caused in the function of EDL to elevate the toes. For that a fast development of force is necessary at the insertion. This can be realized by the high glycolytic activity of fibres in this region. In contrast, the fibres of SOL, characterized as endurance muscle with predominant oxidative metabolism, show no variations in glycolytic activity from the origin to the insertion. However, the oxidative (SDH) activity varies along the longitudinal axis of muscle; it is highest in the middle part of the muscle. This suggests that the highest performance of SOL is there. The variation of SDH activity along the muscle fibre was also found by Edgerton et al. (1990).

Age-dependent changes of oxidative and glycolytic activities of a given fibre type vary in different muscles as shown in Sect. 5.2. However, increasing ATPase activity in all fibre types up to the 21st day after birth occurs independently of the muscle type. The reason for this is the postnatal development of all rat skeletal muscles up to the 21st day, the time for rats to leave the nest. Up to this time the contractility and consequently the ATPase activity of muscle fibres by rights increase in all skeletal muscles.

At the age of 180–200 days the hamsters BIO 8262 show the symptoms of hereditary myopathy. Changes in muscle fibre metabolism were measured only in SO fibres of proximal muscles and not in SO fibres of distal muscles suggesting an effect of this myopathy form mainly on proximal muscles.

Acute hypoxia effects the metabolism of SO and FOG fibres dependent on the metabolic character of the muscle to which they belong. We have shown that both glycolytic and contractile capacity of SO and FOG fibres from fast twitch muscles with predominantly glycolytic metabolism are more variable than of SO and FOG fibres from slow twitch muscles with high oxidative metabolism.

Diabetes results in an diabetic myopathy which is characterized among other things by a fibre atrophy. FG and FOG fibres atrophy to a different degree depending on the muscle type. It is always the fibres of the dominant fibre type of the muscle which atrophy the most, for instance the FG fibres in EDL muscle and the SO fibres in SOL muscle. Moreover, in diabetic rats the FOG fibres of proximal muscles become more oxidative in contrast to FOG fibres of distal muscles which become more glycolytic.

The protective effect of EGb 761 on muscles from hypoxic or diabetic rats is an increase of enzyme activities. The FOG fibres of EDL are more effected than the FOG fibres of SOL. Once more one can state that both the glycolytic and the contractile capacity of SO and FOG fibres from fast twitch muscles with predominantly glycolytic metabolism are more variable than those of SO and FOG fibres from slow twitch muscles with high oxidative metabolism.

In conclusion: Statements about the properties including the adaptability of a given fibre type are actually valid only for that muscle from which the fibres arose. True, a given fibre type is characterized by typical properties. But these properties vary depending on the muscle type they arise from.

10 Correspondences Between Physiological–Metabolic Fibre Typing, ATPase-Fibre Typing and Differentiation of Myosin Isoforms

The cytophotometrical measurements of enzyme activities presented in this study were performed in defined muscle fibres which were physiologically–metabolically classified. Changes in enzyme activities are related to a defined fibre type of this classification system. Besides this physiological–metabolic typing, there is the ATPase-typing which is based on different myosin isoforms in different muscle fibres. As already mentioned in Chap. 3, such ATPase-typed fibres are not clearly metabolically defined. Changes of myosin isoforms, but not of enzyme activities, can be clearly related to ATPase-fibre types. It is a matter of two different classification systems, whose fibre type designations are not changeable in any way. Each of the two systems characterizes specific properties of muscle fibres. Comparing metabolic properties and the content of myosin isoforms of a fibre type, correspondences between the different designations of a fibre type can be detected. It was already shown, see Chap. 3, that SO fibres are identical with the ATPase-fibre type I. In contrast, FOG and FG fibres were not clearly exchangeable with ATPase-fibre types IIA and IIB. However, our own studies and studies of Schiaffino et al. (1990) and Larsson et al (1991) suggest: If we consider the phenomenon of heterogeneity of FOG fibres on the one hand and of heterogeneity of IIB fibres on the other hand, it should be possible to find further correspondences between the two classification systems. In detail, the studies of Schiaffino et al. (1990), Bottinelli and Reggiani (2000) and Larsson et al. (1991) revealed three populations of type II fibres which differ in their heavy chains myosin isoforms (MHC). This means that three fast MHC isoforms were detected on the protein level. Using monoclonal antibodies against these three MHCs, the fibre types IIA, IIB and IIX were identified immunohistochemically. It was pointed out that the IIB fibres which were as usual typed after ATPase reaction following acid preincubation include two distinct fibre populations: "true" IIB and IIX fibres. These two types differ in their SDH activity. "True" IIB fibres showed low SDH activity; IIX fibres showed moderate or high SDH activity. These results correspond to our findings on serial sections (see Fig. 2). We have shown that some IIB fibres (fibres 4, 5) are FG fibres with low SDH activity, whereas other IIBfibres (fibres 2, 3) are FOGI fibres with moderate SDH activity. This means that the "true" IIB fibres defined by Schiaffino (1990) could be FG fibres, and IIX fibres may be identical to FOGI fibres. The other part of FOG fibres, the FOGII fibres, are evidently IIA fibres as shown on serial sections (fibres 6, 7, 8 in Fig. 2).

The classification of muscle fibres is also made complicated by the possible existence of several regular combinations of MHC isoforms in one and the same muscle fibre. For example, in the fibres of rat soleus muscle the coexistence of I- and IIA-MHC, of IIA- and IIX-MHC and of I-, IIA- and IIX-MHC was detected (Schiaffino 1990). IIX-MHC is the MHC isoform with the third fastest electrophoretical mobility, also called IID- MHC (Termin et al. 1990). In this way, the coexistence of IIA and IIX may be reflected in the incompletely separated band IIA/D after gelelectrophoresis of soleus and extensor digitorum longus muscles (Fig. 38).

The expression of multigen families of myofibrillic proteins is regulated on the one hand by development and on the other hand by exogeneous factors, such as use, non-use, motoneuron activity, oxygen supply and hormones (Pette 1990b). By this, several isoforms of proteins or their subunits can be coexistent in one muscle fibre. In this way, a scope for molecular and functional properties which enables the adaptation of muscle fibres to changed functional demands exists. The adaptation can result in fibre type transformations from fast to slow as described in Chap. 6. Muscle fibres are not static entities; they have a "dynamic nature" (Pette 1990b).

In conclusion, with only one classification system alone, not all aspects of fibre alterations under changed conditions are recognizable. The parallel application of physiological–metabolic fibre typing, ATPase-fibre typing and the differentiation of myosin isoforms leads to real muscle analysis. Considering the heterogeneity of FOG-fibres and the heterogeneity of IIB fibres, new correspondences between fibre type designations from different classification systems could be detected.

11 Expression of Nitric Oxide Synthase Isoforms and Protein Kinase-Cθ in the Different Fibre Types and Alterations by Diabetes and EGb 761 Pretreatment

11.1
Illustration of the Topic

Nitric oxide (NO) produced by nitric oxide synthase (NOS) from l-arginine is a physiological modulator of skeletal muscle function and is involved in force development (Kobzik et al. 1994), oxidative metabolism (Kobzik et al. 1995) and regulation of glucose transport (Balon and Nadler 1994; Bedard et al. 1997; Kapur et al. 1997; Reid 1998). Several studies revealed that all three isoforms – NOSI (neuronal NOS), NOSII (inducible NOS) and NOSIII (endothelial NOS) – are expressed in skeletal muscle fibres (Kobzik et al. 1994 1995; Förstermann and Kleinert 1995; Park et al. 1996; Christova et al. 1997; Grozdanovic et al. 1997; Reid 1998; Gath et al.1996, 1999) as well as in the myocard (for review see Buchwalow et al. 2001). Individual muscle fibres can express one or more NOS isoforms (e.g. Reid 1998). However, the expression of the NOS isoforms in skeletal muscle fibres still remains a matter of debate. The sensitivity of the used immunohistochemical technique could be a factor which effects the demonstration of an isoform in histological sections. For instance, there are contradictionary reports about whether the isoform NOSII can be expressed constitutively or only in diseased states (for review see Förstermann and Kleinert 1995; Gath et al. 1996, 1999; Park et al. 1996; Buchwalow et al. 2001). Generally, NOSII was found to be inducible in macrophages by endotoxin and cytokines (for review see Förstermann and Kleinert 1995). Constitutive NOSII expression was detected in human skeletal muscles (Park et al. 1996), in skeletal muscles from guinea pigs (Gath et al. 1996), and in mouse skeletal muscles (Gath et al. 1999). To answer the question whether the NOSII expression is constitutive or not in skeletal muscle fibres, further studies are necessary.

The correlation between the expression of NOS- isoforms and the ATPase-fibre types I and II was investigated for the first time by Kobzik et al. (1994, 1995). They found a correlation of NOSI to type II-fibres in rat muscles, while NOSIII did not correlate to ATPase-fibre types. However, NOSIII was co-expressed with mitochondrial markers. Here it seems useful, but also laborious, to correlate the expression of NOS isoforms rather to the metabolic fibre types SO (slow-oxidative), FOG (fast-oxidative glycolytic), and FG (fast-glycolytic) than to the ATPase-types I and II.

Furthermore, it is known (Rotenberg 1999) that the NO-level in myocytes is influenced by the protein kinase-C (PKC). PKCθ is the most abundant isoform of PKC in skeletal muscle and has been implicated in either the regulation of NOS activity or its expression (Rotenberg 1999). The correlation of PKCθ to NOS isoforms in a given fibre type should be of interest.

As mentioned above, NO can modulate glucose metabolism in skeletal muscles. Therefore, the NOS expression should be altered in muscles of diabetic rats where the glucose uptake in the myocytes is diminished. In Sect. 5.5.3, we have shown the protective effect of Ginkgo biloba extract (EGb 761) on several enzymes of diabetic muscles. In this respect, the effect of EGb 761 on the expression of NOS isoforms in diabetic muscles might be of therapeutic importance.

In the following we will demonstrate immunohistochemically the NOS isoforms I and III and PKCθ in the vastus lateralis muscle of rat. We revealed a correlation in the NOS and PKCθ expression pattern in SO, FOG and FG fibres (Punkt et al. 2001). Additionally, we compared the expression of NOS isoforms by Western blotting in vastus lateralis muscles from different experimental groups: normal and diabetic rats, with and without pretreatment of EGb 761. Finally, we will deal with the problem of NOSII expression in skeletal muscles.

11.2
Experimental Basis

The investigations were performed on rats from 4 experimental groups as described in Sect. 5.5.1. In brief: group 1, normal: 5 rats without any special treatment served as controls. Group 2, diabetic: Diabetes was induced by intraperitoneal administration of 60 mg/kg body weight streptozotocin (STZ; Boehringer, Mannheim, Germany). Group 3: Normal rats were treated with Ginkgo biloba extract (EGb 761, IPSEN, Paris, France). Group 4: Diabetic rats were treated with EGb 761. Animals were anaesthetized with ether and then decapitated. The vastus lateralis muscle was removed; muscle samples from the middle portion between origin and insertion of the muscle were prepared, powdered with talcum and frozen in liquid nitrogen. 10-μm thick sections of normal vastus lateralis muscle were made for enzymehistochemistry and immunohistochemistry. A cryostat 1800 (Reichert Jung, Wien, Austria) was used.

Enzyme Histochemistry. Activity of succinate dehydrogenase (SDH; E.C. 1.3.5.1) was demonstrated according to Lojda et al. (1976). The method as described by Lojda et al. (1976) to demonstrate activity of mitochondrial glycerol-3-phosphate dehydrogenase (GPDH; E.C. 1.1.99.5) was modified: the medium consisted of 0.1 M phosphate buffer, pH 7.4, 5 mM DL-3-glycerophosphate, disodium salt, 0.01% menadione and 0.5 mg/ml nitro blue tetrazolium chloride (NBT). The incubation period was 20 min at 37°C. The activity of myofibrilic adenosine triphosphatase (ATPase; E.C. 3.6.1. 32) was demonstrated according to Padykula and Herman (1955) after preincubation at pH 9.4.

Control reactions in the absence of substrate were performed each time on serial sections.

Fibre Typing. Fibre typing into SO (slow-oxidative), FOG (fast-oxidative glycolytic) and FG (fast-glycolytic) has been described (Punkt et al.,1996 1998) based on cytophotometrically determined activities of GPDH, SDH and ATPase in one fibre in serial cross sections. FOG fibres were subdivided into FOGI fibres with low SDH and moderate GPDH activity and FOGII fibres with high SDH and low GPDH activity.

Immunohistochemistry. Cryosections were fixed in cold acetone for 15 min and thoroughly air-dried for 30 min. After rinsing with phosphate-buffered saline (PBS), non-specific binding sites were blocked with PBS containing 10% goat serum for 30 min. PBS was also used for all washes and dilutions of antibodies (AB). The cryosections were immunoreacted overnight at 4°C with rabbit primary polyclonal AB recognizing NOSI, NOSIII and polyclonal or monoclonal AB against PKCθ (Transduction Laboratories). Primary AB were diluted to a final concentration of 2.5 μg/ml. To quench endogeneous peroxidase activity, sections were treated with methanol containing 1.2% H_2O_2 for 15 min. Bound primary AB were detected employing Cy3 anti-mouse (Jackson Immuno Research Laboratories, Inc.) and HRP (horseradish peroxidase)-conjugated goat anti-rabbit secondary AB with following FITC-tyramide signal amplification.

The controls were: (a) omission of primary AB; and (b) substitution of primary AB by rabbit IgG (Dianova) at the same final concentration as primary AB.

Western Blotting. Samples from all experimental groups were investigated. Total protein of samples harvested in PBS and sedimented by centrifugation was homogenized in 10 mM Hepes, pH 7.5, 0.2 mM phenylmethylsulfonylfluoride (PMSF) and 0.1 mM dithiothreitol (DTT) with Ultra Turrax at 30 000 rpm for 3×5 s. The final pellet was suspended in 0.75 ml of 10 mM Hepes, pH 7.5 250 mM sucrose, 0.2 mM PMSF and 0.1 mM DTT and stored at –80°C until use. 40 μg protein samples solubilized in SDS sample buffer were separated by 7.5% Laemmli-polyacrylamide gel electrophoresis (Laemmli 1970). Separated proteins were electrotransferred onto polyvinylidenedifluoride membranes. Processing for immunoblotting was performed as described in Towbin et al. (1979). Dilution of primary anti-NOS antibodies was performed according to the manufacturer's instructions (Santa-Cruz and Transduction Laboratories). The secondary antibody was peroxidase-labeled anti-rabbit IgG (Sigma 1:15,000). The immunoreaction was visualized using ECL-Kit (Amersham).

11.3
The Correlation of NOSI, NOSIII and PKCθ to the Fibre Types

The analysis of serial muscle sections of vastus lateralis muscle provided a correlation matrix (Table 5) of the activities ATPase, SDH, GPDH and the immunoreactivities of PKCθ, NOSI and NOSIII in the different fibre types (Fig. 39). NOSI, NOSIII and PKCθ expression patterns were fibre type specific. Fibre typing on serial sections revealed the type of positive fibres. Both NOSI and PKCθ immunoreactivity were observed throughout the cytoplasm and at the sarcolemma of SO and FOGII fibres. These are fibres with predominantly oxidative metabolism. The fibres, which were positive for

Table 5. Correlation of the immunoreactivities of PKCθ, NOSI, NOSIII and the activities of SDH, GPDH, ATPase and in the different fibre types of rat vastus lateralis muscle

ATPase-fibre type	Metabolic fibre type	PKCθ	NOSI	NOSIII	SDH	GPDH	ATPase
I	SO	+	(+)	–	+	–	–
II	FOGI	-	(+)	+	(+)	+	+
II	FOGII	+	+	-	+	-	(+)
II	FG	–	(+)	+	–	+	+

+ strong staining; (+) weak staining; – no or very weak staining.

both NOSI and PKCθ, showed at the same time high SDH activity (see Figs. 39a,e,d). FG and FOGI fibres showed weak NOSI immunostaining and were PKCθ-negative. NOSIII immunoreactivity was found in fibres with predominantly glycolytic metabolism, FOGI and FG. NOSIII immunoreaction correlates to GPDH activity (Figs. 39c and f). Fibres with positive NOSIII immunoreaction (FOGI, FG) were negative for PKCθ (Figs. 39c,e). However, they show a weak immunostaining for NOSI (Figs. 39c and a). FOGII and SO fibres are NOSIII-negative.

The immunoreaction of NOSI as well as of NOSIII isoforms was found as diffuse or granular staining throughout the cytoplasm and at the sarcolemma of defined muscle fibres. Kobzik et al. (1995) found similar NOSIII-staining. Frandsen et al. (1996) investigated human skeletal muscles and found immunoreactivity of NOSI in the sarcolemma and the cytoplasm of all muscle fibres. Type I-muscle fibres expressed stronger NOSI immunoreactivity. Moreover, the histochemical staining for cytochrome oxidase showed a similar staining pattern to that of NOSI immunoreactivity. These results agree with our findings, showing that NOSI correlated with SDH activity. Several studies revealed that NOSI was associated to the sarcolemma of rat skeletal muscle fibres (for references see Christova et al. 1997; Grozdanovic et al. 1997).

We used for this and recent studies an advanced immunocytochemical method of signal amplification with fluorescein-labeled tyramine. Buchwalow et al. (1996, 1997, 1997a, 2001) have shown that NOSII as well as NOSI are associated with mitochondria, with contractile fibres, along the plasma membrane and T-tubules of cardiomyocytes. Our finding that NOSI is localized not only at the sarcolemma but also in the cytoplasm (apparently including mitochondria) of skeletal muscle fibres may be due to the improved technique that we have used and supports the results in cardiomyocytes (Buchwalow 2001). Assuming, that the immunolabeling characterizes the expression of the enzyme, the following conclusions can be made: NOSI and PKCθ are co-expressed in fibres with predominantly oxidative metabolism (SO, FOGII). This suggests that PKCθ is involved in the regulation or the expression of the isoform NOSI to influence the NO level in oxidative fibres. Kobzik et al. (1994) showed that high NO levels diminished the skeletal muscle contraction, inhibiting force output. One could conclude that the force output of oxidative fibres may be regulated by the interplay of

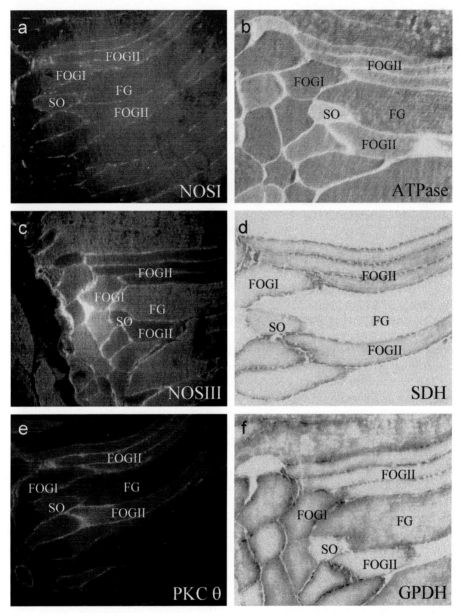

Fig. 39. Immunoreactivities of *NOS I* (**a**), *NOS III* (**c**), *PKCθ* (**e**) and activities of *ATPase* (**b**), *SDH* (**d**), *GPDH* (**f**) of the fibre types *SO, FOGI, FOGII* and *FG* detected in cryostat serial sections of rat vastus lateralis muscle. × 100

NOSI and PKCθ. In contrast to oxidative fibres, glycolytic fibres (FG) showed no PKCθ expression suggesting that the NOS expression of this fibre type is not effected by PKCθ. We found NOSIII expression in fibres with predominantly glycolytic metabolism (FOGI and FG) and not in oxidative fibres. Additionally, a weak NOSI expression was shown in FOGI and FG fibres. This means that FOGI and FG fibres can co-express NOSI and NOSIII. However, SO and FOGII fibres, which co-express NOSI and PKCθ, do not express NOSIII. In other words, NOSI is co-expressed either with PKCθ (in SO and FOGII fibres) or with NOSIII (in FOGI and FG fibres).

Kobzik et al. (1995) investigated the correlation of NOSI and NOSIII to the ATPase-fibre types I and II. They found NOSI in type II-fibres. These fibres include the FOGII fibres, which were found to be NOSI-positive in the present study. Kobzik et al. (1995) showed that some type II-fibres co-expressed NOSI and NOSIII. These fibres may be FG and FOGI fibres as we have shown. Kobzik et al. (1995) found no correlation of NOSIII to ATPase-fibre types, but NOSIII correlated positively with the SDH activity as a mitochondrial marker in extensor digitorum longus muscle and diaphragm. Our findings in vastus lateralis muscle revealed that NOSIII correlated to the mitochondrial marker GPDH, whereas an inverse correlation with SDH activity was observed. We assume that the fibre type specific expression of NOSIII may vary in different types of muscles.

11.4
The Effect of STZ-Induced Diabetes and EGb 761 on the Expression of NOSI and NOSIII in the Rat Vastus Lateralis Muscle

Differences in both NOSI and NOSIII expression between muscles from different experimental groups were detected on the protein level by Western blotting (Figs. 40, 41). The results were similar for NOSI and NOSIII. The interpretation of blots is as follows: The NOSI expression of vastus lateralis muscle decreased in the following order of experimental groups: EGb 761-treated normal rats >EGb 761-treated diabetic rats >normal rats >diabetic rats. In the case of NOSIII, the order from high to low expression was: EGb 761-treated normal rats >normal rats >EGb 761-treated diabetic rats >diabetic rats. The NOSIII expression in normal muscle was low, that of NOSI very low, and both isoforms disappeared in diabetic muscle. Treatment of rats with EGb 761 resulted in an increase of NOSI as well as of NOSIII expression. Muscles from normal, EGb 761-treated rats showed the highest NOSI and NOSIII expression of all experimental groups. In contrast to muscles from untreated diabetic rats, muscles from EGb 761-treated diabetic rats expressed NOSI and NOSIII. The NOSI and NOSIII expression of EGb 761-treated diabetic muscles was lower than that of EGb 761-treated normal muscles. Summarizing, diabetes decreased, and EGb 761 treatment increased the NOSI as well as the NOSIII expressions in vastus lateralis muscle.

The decrease or disappearance of NOSI expression in diabetic muscles may be a response to diabetic neuropathy. It is known that diabetes induces neuropathy with damaged peripheral nerves. Grozdanovic et al. (1997) found neuronal effects on skeletal muscle NOSI. Moreover, several authors (Balon and Nadler 1994; Bedard et al. 1997; Kapur et al. 1997; Reid 1998) showed, that basal glucose metabolism in rat skeletal muscles are regulated by NO. They demonstrated that NOS activity is correlated with the glucose uptake in skeletal muscles. Inhibition of NOS decreased the glucose uptake. The uptake of glucose was shown to be reduced in diabetic muscles. These considerations are well correlated with our findings, that NOSI and NOSIII were down-regulated in diabetic muscles.

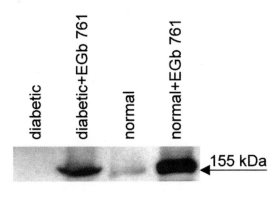

NOS I

Fig. 40. Western blotting of NOS I in vastus lateralis muscles of rats from four experimental groups: diabetic rats *(diabetic)*, diabetic rats treated with Gingko biloba extract *(diabetic + EGb 761)*, normal rats *(normal)*, normal rats treated with Ginkgo biloba extract *(normal + EGb 761)*

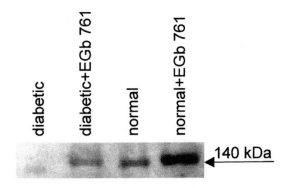

NOS III

Fig. 41. Western blottting of NOS III in vastus lateralis muscles of rats from 4 experimental groups: diabetic rats *(diabetic)*, diabetic rats treated with Gingko biloba extract *(diabetic + EGb 761)*, normal rats *(normal)*, normal rats treated with Ginkgo biloba extract *(normal + EGb 761)*

Furthermore, an enhanced NOSI and NOSIII expressions in normal and diabetic muscles after treatment of rats with EGb 761 suggests that glucose uptake may be improved by EGb 761. We have demonstrated a protective effect of EGb 761 on enzyme activities of diabetic skeletal muscle (see Sect. 5.5.3). Evidently, EGb 761 protection of diabetic skeletal muscles is manifold.

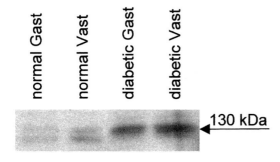

NOS II

Fig. 42. Western blotting of NOSII in vastus lateralis *(Vast)* and gastrocnemius *(Gast)* muscles of normal and diabetic rats

11.5
NOSII Expression in Skeletal Muscle Fibres

Recently, we found NOSII immunostaining in the fibres of vastus lateralis muscle and in the capillaries of gastrocnemius muscle of normal rats. The immunoreaction of NOSII seems to differ between different muscles: while the NOSII immunoreaction of vastus lateralis muscle was found diffusely distributed in all fibres, the fibres of gastrocnemius muscle showed no staining. However, in contrast to vastus lateralis muscle, the capillaries of gastrocnemius muscle were NOSII-positive. The analysis of Western blots (Fig. 42) revealed a weak NOSII expression in both muscles, suggesting NOSII may be expressed constitutively. Moreover, the NOSII expression was elevated in both muscles of diabetic rats (Fig. 42). This result may confirm findings from literature. Bedard et al. (1997) showed that glucose transport and insulin action in skeletal muscle cells were affected by NOSII expression and NO production. Kapur et al. (1997) found that NOSII induction, concomitant with elevated NO production, is associated with impaired insulin-stimulated glucose uptake in isolated rat muscles.

However, additional studies about NOSII expression in skeletal muscles are necessary to investigate which fibre types express NOSII and in which way the NOSII expression depends on muscle type and on the species.

11.6
Conclusions

Our data demonstrate for the first time a different expression pattern of NOS isoforms I and III and PKCθ in the physiological–metabolic fibre types SO, FOG and FG. Generally, NOSI and PKCθ were co-expressed in fibres with predominantly oxidative metabolism (SO, FOGII). This suggests an interplay of PKCθ and NOSI in nitric oxide production by oxidative fibres. NOSIII was more expressed in fibres with predominantly glycolytic metabolism (FOGI, FG). Weak NOSI immunoreactivity was also found in NOSIII-positive fibres suggesting that NOSIII and NOSI are co-expressed in these fibres. Western blotting revealed that NOSI as well as NOSIII expression in the vastus lateralis muscle was decreased or disappeared in diabetes and increased by Ginkgo biloba extract treatment. These effects may be associated with a diminished glucose uptake by myocytes of diabetic muscles and to an improved glucose uptake after Ginkgo biloba treatment.

NOSII seems to be constitutively expressed in vastus lateralis and gastrocnemius muscles. Elevated expression of NOSII was shown in diabetic muscles. This is supported by findings from literature showing that NOSII is induced in muscles associated with impaired glucose uptake.

12 Specific Fibre Types of Extraocular Muscles

KARLA PUNKT, GERHARD ASMUSSEN[1]

The previous chapters of this book deal with the fibre types of vertebrate hind limb muscles which represent typical skeletal muscles. Several other cross striated muscles differ from the typical skeletal muscles in their MHC gene expression and functional specialization. Extraocular muscles are such "atypical" muscles. Their fibre types do not correspond to those of skeletal muscles; e.g. they possess slow muscle fibres that are more characteristic of avian and amphibian muscles. In the following an overview about the functional and structural organization of extraocular muscles and their unusual fibre types is given.

12.1
The Complexity of Extraocular Muscles

The six rotatory extraocular muscles (EOMs) are structurally and functionally unique among the cross-striated muscles of vertebrates. They exhibit a diverse repertoire of actions including steady eyeball fixation, slow vergence movements, pursuit movements at various speeds, and high-speed saccades over a wide range of angles (Leigh and Zee 1991). In many mammalian species EOMs are characterized by a much more rapid time course of isometric twitch contractions, a higher fusion frequency, a lower twitch-to-tetanus ratio combined with a higher fatigue resistance, caused by a high oxidative as well as a high glycolytic capacity, than fast-twitch muscles of the same individual (for review see Asmussen and Gaunitz 1981a). In comparison to fast-twitch limb muscle, they also show a low force output and an exceptionally high shortening velocity, independent of the size of the animal (Asmussen et al. 1994). The investigation of the cross-bridge kinetics of single glycerinated fibres of EOMs of the rabbit also shows the existence of superfast twitching fibres (Li et al. 2000). Besides these fast-twitch fibres the EOMs also contain a minority of multiply-innervated, slow-tonic muscle fibres – very rare generally in mammalian muscles – which contract extremely slowly (for review see Morgan and Proske 1984). This tonic fibre system is possibly responsible for the fact that EOMs display some properties normally observed only in muscles of lower vertebrates or in neonatal or denervated

[1] Gerhard Asmussen, Carl-Ludwig-Institute of Physiology, University Leipzig, Liebigstr. 27, 04103 Leipzig, Germany

mammalian muscles, e.g. acetylcholine contractures (Asmussen and Gaunitz 1981b). EOM also possess some unusual structural features: they contain very thin fibres in a layer organization, first described by Kato (1938).Many fibres branch and form muscle-to-muscle junctions (Mayr et al., 1975), and their motor unit size is very small (Torre 1953).

These functional and structural intricacies are reflected in the complex muscle fibre types described in mammalian EOMs. Their fibres have been classified into six different types on the basis of their arrangement in the muscle, their enzyme and immunohistochemistry, ultrastructure, and innervation (see below). This system of classification differs markedly from that normally used to classify functionally different limb muscle fibres, subserving the posture and locomotion of the animal. Therefore, the fibre types in EOMs do not correspond to those in skeletal muscles (Porter et al., 1995). Limb muscle fibres are subdivided into four types based on myosin isoforms: type I or slow fibres and three subtypes of type II or fast (IIA, IIB, IIX or D) fibres, each expressing a different isoform of myosin heavy chain (MHC). Differences in metabolic properties, myosin-ATPase activity, shortening velocity, force generation, and thermodynamic efficiency observed between the individual fibre types are directly correlated with the presence of specific MHC isoforms (for review see Schiaffino and Reggiani, 1996). An indicator of the complexity of the EOMs is seen in the fact that eight different MHCs have been identified in these muscles in adult mammals. These include isoforms found in adult limb fast (type IIA-, IIX-, IIB-MHCs) and slow (type I-MHC, identical with the cardiac-β-MHC) fibres (Asmussen et al. 1993; Wieczorek et al. 1985), but in addition, MHCs found in developing (embryonic and fetal/neonatal-MHCs) but not in mature limb muscles (Wieczorek et al. 1985; Asmussen et al. 1993), the cardiac-specific α-MHC (Pedrosa-Domelöff et al. 1992; Rushbrook et al. 1994), as well as an EOM-specific isoform, the EOM-MHC (Sartore et al. 1987, Lucas et al. 1995), and the slow-tonic-MHC (Pierobon-Bormioli et al. 1995), an isoform in mammals normally expressed only in the nuclear bag fibres of muscle spindles.

After this more functional implication, we will now describe more in detail the fibre types of extraocular muscles. In the early seventies of the last century, three research groups independently suggested a differentiation into six different fibre types in mammalian EOMs (Asmussen et al. 1971, cat and rabbit; Mayr 1971, rat; Harker 1972, sheep). In the meantime, this classification has often been confirmed and now is generally accepted (for review see Spencer and Porter 1988).

12.2
Fibre Types of Extraocular Muscles

EOMs exhibit two distinct regions: an orbital layer directed to the bony orbit and a global layer faced to the eyeball. Each of the two regions comprise specific fibre types. This means that besides other criteria the fibre types of EOMs are also determined by their location. The transition between the orbital and global layers is the intermediate zone which contains an admixture of fibre types from either zone. Table 6 gives an overview about the properties of the six fibre types in EOMs on the basis of location, histochemistry and pattern of innervation, ultrastructure and MHC isoforms. The fibre types are characterized as follows:

Table 6. Morphological properties of extraocular muscle fibre types

	Orbital		Global			
Fibre type	1	2	3	4	5	6

Histochemistry (Spencer and Porter 1988, Rowlerson 1987, and own observations)

Diameter (µm)	25±4	19±3	27±5	35±5	47±6	36±4
Percentage	80%	20%	33%	25%	32%	10%
ATPase 9.4	+++	+/++	+++	+++	+++	–/+
ATPase 4.3	+/–	+++	+/–	+/–	+/–	+++
SDH	++++	++	++++	+++	++	+/–
GPDH	+	+	+++	+++	+++	+/–
AChE	Focal	Multiple	Focal	Focal	Focal	Multiple

Ultrastructure (Mayr 1973)

Myofibrils

Extent	Small	Large	Small	Small	Small	Large
Volume	60%	78%	55%	65%	71%	83%
SR (volume)	9%	6%	10%	14%	16%	4%
Mitochondriax (volume)	20%	6%	24%	13%	5%	5%
Z line (width in nm)	73	118	76	54	48	100

MHC Isoforms (Rubinstein and Hoh 2000, and own observations)

	Embryonic	Embryonic	IIA	IIX	IIB	I
	EOM	I Slow-tonic (some) a-cardiac (some)				

Abbreviations: ATPase, myofibrillic adenosine triphosphatase; SDH, succinate dehydrogenase; GPDH, glycerol-3-phosphate dehydrogenase; AChE, acetylcholinesterase (staining of motor endings); SR, sarcoplasmatic reticulum; MHC, myosin heavy chain; ++++ very strong, +++ strong, ++ moderate, + weak, – negative.

The orbital layer contains fibre types 1 and 2. Fibre type 1 is with 80% the predominant fibre type in the orbital layer. These fibres are single innervated, demonstrated by focal acetylcholinesterase staining of motor endings. The strong staining of alkaline stabile ATPase activity suggests that this fibre type may be physiologically fast twitch. Strong staining for SDH activity and weak staining for GPDH activity were found. Fibre type 2 comprises the orbital multiple innervated muscle fibres revealed by multiple AChE localization. The pattern of myosin ATPase activity is similar to that of nuclear bag intrafusal fibres. The intensity of staining for SDH is moderate, that for GPDH is weak. Ultrastructural differences between fibre type 1 and 2 are mainly the content of mitochondria, which is three times higher in type 1 fibres. However, the ultrastructure of both fibre types vary over the fibre length. For example, fibres of type 2 display twitch-like ultrastructural features in the vicinity of the end plate region, while the proximal and distal regions exhibit tonic-like features (for references see Spencer and Porter 1988).

The global layer contains three fibre types (3, 4, 5) of single innervated fibres and another type (6) of multiple innervated fibre. The fibre types 3, 4, 5 may differ in their diameter and ultrastructure. However, the ATPase, SDH and GPDH activities are quite similar between these fibre types. Many fibres do not fit the criteria of a single fibre type, suggesting that the single innervated fibre types of the global layer form a continuum of fast twitch fibres (Nelson et al. 1986). Similar to the FOG fibres of skeletal muscle fibres, the EOM global fibre types 3, 4, 5 show both high SDH and high GPDH activities. The multiple innervated fibre type 6 shows very weak staining for alkaline stabile ATPase, SDH and GPDH. This fibre type contains very few and small mitochondria. The myofibrils are large; the SR is poorly developed, and the extend of the Z line is characteristically wide. The histochemical as well as the ultrastructural profile is similar to that of slow-tonic muscle fibres in amphibians (for references see Spencer and Porter 1988). These fibres physiologically exhibit a slow-graded, non-propogated response following neural or pharmacological activation.

Up to now, the metabolic differentiation of the EOM fibre types was based on the rough subjective evaluation of the enzyme reactions in the fibres. For objective results quantitative enzyme histochemistry is necessary. Our current investigations deal with cytophotometrical studies of EOM fibres to reveal the metabolic characteristic of the different fibre types.

The distribution of myosin heavy chain (MHC) isoforms among the rat EOM fibre types was investigated recently by Rubinstein and Hoh (2000). The EOMs of adult rats express mRNAs for at least eight distinct MHC isoforms, mentioned above. The MHC isoforms were immunohistochemically localized within adult EOM fibres. It was shown that each orbital fibre synthesize more MHC isoforms which are localized differently along the length of the fibre. EOM-specific MHC is expressed in the endplate region, while the embryonic MHC is excluded from the endplate region but is present in the rest of the fibre. In contrast to orbital fibre types, the global fibre types synthesize predominantly one specific MHC isoform without longitudinal variation. According to the expression of the different MHC isoforms, two orbital and four global fibre types were differentiated. The correlation of the MHCs and the histochemically defined fibre types is given in Table 6.

12.3
Summary

Extraocular muscles are complex in their structure and function. They exhibit a considerably greater variability of MHCs than limb muscles. In addition to adult MHC isoforms found in skeletal muscle, they synthesize a slow tonic MHC, α-cardiac MHC, an EOM-specific MHC and MHC isoforms normally found only in developing skeletal muscle fibres. The expression of one or more specific MHC isoforms in each fibre determines six different fibre types of extraocular muscles. These fibre types differ from usual skeletal muscle fibre types. Two fibre types (type 2 and 6) comprise multiple innervated fibres unseen in skeletal muscles. Cytophotometrical measurements can clear up the metabolic profile of the EOM fibre types.

13   Summary

Fibre diversity is an intrinsic property of skeletal muscle. The "knowledge" of a muscle fibre about whether it has to be slow or fast derives from the genetic code. During their lifetime the fibres do not forget their inaugural program, but they can adapt to changed functional conditions by varying their properties. The aim of this review was to characterize the fibre types of rat skeletal muscles from different points of view and to investigate the adaptability of fibre types to altered physiological and pathological conditions. An overview about the different classification systems and their underlying criteria has been given. Studies from literature as well as our own studies revealed that with only one classification system not all aspects of fibre alterations under changed conditions are recognizable. The parallel application of ATPase-fibre typing, the differentiation of myosin isoforms and metabolic fibre typing leads to real muscle analysis. It was shown that the metabolic situation of a given fibre type in heterogeneously composed muscles can be characterized by cytophotometry. This is a method of quantitative histochemistry based on absorbance measurements of the final reaction product of an enzyme reaction in the histological section. The reliability of cytophotometrical data was proved by the strong correlation to biochemically measured enzyme activities. Fibre typing by cytophotometry has been established and described in detail in the present book. The muscle fibres were typed by measurements of the following three enzyme activities in one and the same fibre: myofibrillic adenosine triphosphatase activity as a marker of contractility, glycerol-3-phosphatedehydrogenase as a marker of glycolysis and succinate dehydrogenase as a marker of oxidative metabolism. The resultant physiological-metabolic fibre types were SO (slow-oxidative), FOG I (fast-oxidative glycolytic fibre type with moderate SDH activity), FOG II (fast-oxidative glycolytic fibre type with high SDH activity) and FG (fast-glycolytic).

Cytophotometrical measurements of the three chosen enzyme activities provided changes in the metabolic profile of physiological–metabolic fibre types. Extensive cytophotometrical and morphometrical analyses of several rat hind limb muscles were performed to demonstrate changes of muscle fibre properties under physiological and pathological conditions. Regional differences within the muscles, effects of ageing, myopathy, hypoxia, diabetes and Ginkgo biloba treatment on the different fibre types were investigated. From these investigations, statements about properties including adaptability of a given fibre type were possible. FOG fibres proved to be the most adaptable muscle fibres. The fibre type properties vary depending on the muscle type the fibres arise from. Summarizing the large number of results, one can generalize and make the following statements:

- Muscle fibres adapt themselves to physical demands in different parts of the muscle by varying their size and enzyme activities. The adaptation varied dependent on fibre type and muscle type and need not be uniform throughout the whole muscle. To investigate changes in muscle fibre properties due to biological or pathological changes, bioptates from identical muscle regions must be taken.
- At the time of birth the differentiation into fibre types is not completed. At the age of 21 days a shift in fibre type population from slow to fast was found in rat hind limb muscles. At this time the metabolic fibre types SO, FOG and FG are distinguishable. All fibre types showed changes in their metabolic profile during postnatal development and ageing. Generally, a fibre type specific increase of enzyme activities was observed up to the 21st day after birth, the time of weaning. After that, the enzyme activities take their adult level which is usually lower than that of the 21st day. During the postnatal development the muscle fibres become more glycolytic. During the adult period of 56–200 days the metabolic profile of all fibre types is constant. In muscles of elder rats with an age of 370 and 550 days further enzyme activity changes of fibre types were measured.
- The metabolic profile of a fibre type differed between normal and myopathic hamster muscles. The direction of metabolic shift as well as the most effected fibre type were dependent on the age of the hamsters. At the time of onset of clinical symptoms, only the metabolic profile of SO fibres differed between normal and myopathic muscles. Hereditary myopathy effected the fibre metabolism in proximal but not in distal muscles.
- After acute short hypoxia, a fibre type-specific increase of enzyme activities and a shift to higher oxidative metabolism were found. These alterations were dependent on the muscle type. Mainly the SO and FOG fibres were affected.
- Streptozotocin-induced diabetes drastically affected the fibre properties. Fibre type-related enzyme activities, fibre type distribution, fibre cross areas and myosin isoforms changed in muscles of diabetic rats. FOG fibres showed a shift to more glycolytic metabolism and about a third of them transformed into FG fibres. SO fibres became more oxidative. Different fibre types atrophied to a different degree. A reduced amount of slow and an increased amount of fast myosin heavy chain isoforms were found.
- After treatment of rats with Ginkgo biloba mainly the oxidative fibres SO and FOG were protected against hypoxic and diabetic alterations.
- Fibre type transformation is a common phenomenon in skeletal muscles under changing conditions. Our results suggest that transformations between FOG and FG fibres occurred. We conclude that FOG and FG fibres represent the dynamic metabolic situation of the fast fibres. Transitions between the physiological slow and fast fibre types as well as between the ATPase-fibre types IIA and IIB are described in the literature. Muscle fibres are not stable entities, rather they have a dynamic nature.
- It was proved that the muscle fibre plasticity represents an important adaptive mechansim.
- In addition to their metabolic characteristic, the physiological-metabolic fibre types were immunohistochemically characterized by their expression of nitric oxide synthase (NOS) isoforms I, III and protein kinase Cθ (PKCθ). This means that for the first time the expression of NOS isoforms and PKCθ was correlated to the physiological–metabolic fibre types. PKCθ and NOSI were co-expressed in

fibres with predominantly oxidative metabolism (SO, FOGII), suggesting an interplay of PKCθ and NOSI in NO production by oxidative muscle fibres. NOSIII and NOSI were co-expressed in fibres with predominantly glycolytic metabolism (FOGI and FG). Moreover, the results suggest a constitutive expression of NOSII in skeletal muscles.

- Finally, specific musle fibre types of extraocular muscles which differ from typical skeletal muscle fibre types were described. Extraocular muscles exhibit a greater complexity of MHCs than other skeletal muscles. In addition to adult MHC isoforms found in skeletal muscles, they synthesize a slow tonic MHC, an EOM-specific MHC and MHC isoforms normally found only in developing skeletal muscle fibres.

14 Conclusions

What are the perspectives for studies of muscle fibre diversity? One point is the investigation of the molecular and functional diversity of human muscle fibres as well as their adaptability and their specific contribution to muscle performance. Reports on fibre diversity so far are predominantly concerned with muscles of animals, mainly of rats and rabbits. These studies revealed that fibre diversity is species-dependent. Recent studies (Bottinelli et al. 2000) suggest that fibre diversity and fibre adaptability in humans differ from that of animals.

Another point is the use of highly specialized techniques of molecular biology in muscle research. In the present review the properties and the adaptability of muscle fibres were described by myofibrillic proteins, the enzyme activities, the morphology and the shortening velocity using "classical" methods. The development of modern molecular biology has provided valuable new insights into muscle fibre research. Today, many processes can be explained at the level of gene expression, e.g. a subset of "fast" genes is expressed when muscle fibres adapt for increased power output (Goldspink 1996). The availability of molecular biology and the knowledge about the human genome sequences opens up new possibilities in medicine in the fields of cardiomyoblasty and the replacement of damaged skeletal muscles. In this respect, the controlled regulation of muscle differentiation is of clinical importance for the development of better treatment strategies. The differentiation of muscle fibres is based on the plasticity of the muscle precursor cells, but it remains to be clarified whether fibre diversity is due to lineage or non-lineage specification. Also other problems are still unsolved. As mentioned in Chap. 6, the genetic regulation of fibre transformation is not yet clear. Present and future investigations have the goal of illuminating the signalling pathway underlying fibre transformation and other adaptation processes. Nowadays, main stream muscle research deals with the intracellular signals which regulate various aspects of muscle function. Findings from these studies will again put new complexions on what is already known, and then muscle fibre diversity can be considered from new perspectives. Here it appears to be suitable to quote from a lecture by Andrew Fielding Huxley, who developed the sliding filament theory. At the 49th meeting of the Nobel Prize winners in Lindau, Germany in 1999, Huxley gave a lecture on the recent advances in muscle research. He described the shift to molecular biology in muscle research as a new paradigm change and as a revolution. Paradigm changes involve a certain risk: to forget the highly specialized knowledge that existed before. Huxley recalled the "revolution" in muscle research around 1900, the result of which he called a catastrophe. Valuable knowledge fell into oblivion because of the change from microscopic to biochemical research. For example, the cross-striation of muscles, discovered in 1876, was forgotten. Not before the middle of the twentieth century was there a renaissance of this finding because of a "counter-revolution," represented by the use of electron microscopy. The change of paradigm here was from biochemistry to ultrastructure. Today, muscle research concentrates more and more on molecular processes. For that reason. there is a danger of "repeating history" 100 years later – meaning that valuable knowledge again could fall into oblivion because of indifference and insufficient interest towards microscopic structures.

According to Huxley's message the perspective for muscle fibre research is as follows: further findings about the different muscle fibres can be gained by applying highly specialized, molecular-biological methods. With the help of this knowledge, already known facts can be evaluated or revised. To my mind, only the integration of classical and modern methods is successful in obtaining new results in muscle fibre research and in applying these results in practical medicine.

References

Aigner S and Pette D (1992) Fast-to-slow transition in myosin heavy chain expression of rabbit muscle fibres induced by chronic low-frequency stimulation. Symp Soc Exp Biol 46: 311–317

Almeida-Silveira MI, Perot C, Goubel F (1996) Neuromuscular adaptions in rats trained by muscle stretch-shortening. Eur J Appl Physiol Occup Physiol 72 (3): 261–266

Andersen JL and Schiaffino S (1997) Mismatch between myosin heavy chain mRNA and protein distribution in human skeletal muscle fibers. Am J Physiol 272: C1881-C1889

Anderson J, Almeida-Silveira MI, Perot C (1999) Reflex and muscular adaptions in rat soleus muscle after hind limb suspension. J Exp Biol 202 (19): 2701–2707

Asmussen G, Gaunitz U (1981a) Mechanical properties of the isolated inferior oblique muscle of the rabbit. Pflügers Arch 392: 183–190

Asmussen G, Gaunitz U (1981b) Contractures in normal and denervated inferior oblique muscle of the rabbit. Pflügers Arch. 392: 191–197

Asmussen G and Soukup T (1991) Arrest of developmental conversion of type II to type I fibres after suspension hypokinesia. Histochem J 23: 312–322

Asmussen G, Kiessling A, Wohlrab F (1971) Histochemische Charakterisierung der verschiedenen Muskelfasertypen in den äusseren Augenmuskeln von Säugetieren. Acta Anat. 79: 526–545

Asmussen G, Traub I, Pette D (1993) Electrophoretic analysis of myosin heavy chain isoform patterns in extraocular muscles of the rat. FEBS Lett. 335: 243–245

Asmussen G, Beckers-Bleuks G, Marechal G (1994) The force-velocity relation of the rabbit inferior oblique muscle; influence of temperature. Pflügers Arch. 426: 542–547

Balon TW, and Nadler JL (1994) Nitric oxide release is present from incubated skeletal muscle preparations. J Appl Physiol 77 (6): 2519–2521

Barker D (1974)The morphology of muscle receptors. In: Hunt CC (edn) Muscle receptors, Handbook of sensory physiology, vol 3. pt 2. Berlin, Springer-Verlag, pp 1–190

Barker D and Banks RW (1994) The muscle spindle. In: Engel AG, Franzini-Armstrong C (eds) Myology. Vol.1. Mc-Graw-Hill Inc, New York, St Louis, San Francisco, Auckland, Bogota, Caracas, Lisbon, London, Madrid, Mexico City, Milan, Montreal, New Dehli, Paris, San Juan, Singapore, Sydney, Tokyo, Toronto, 2nd edn, pp 333–360

Bedard S, Marcotte B, Marette A (1997) Cytokines modulate glucose transport in skeletal muscle by inducing the expression of inducible nitric oxide synthase. Biochem J 325: 487–93

Billeter R, and Hoppeler H (1992) Muscular basis of strength. In: Komi PV (ed) Strength and Power in Sport. Vol III, Blackwell scientific publications, Oxford, London, Edinburgh, Boston, Melbourne, Paris, Berlin, Vienna, pp 39–63

Bischoff R (1994) The satellite cells and muscle regeneration. In: Engel AG, Franzini-Armstrong C (eds): Myology. Vol.1. Mc-Graw-Hill Inc, New York, St Louis, San Francisco, Auckland, Bogota, Caracas, Lisbon, London, Madrid, Mexico City, Milan, Montreal, New Dehli, Paris, San Juan, Singapore, Sydney, Tokyo, Toronto, 2nd edn, pp 97–118)

Blanco CE, Sieck GC, and Edgerton VR (1988) Quantitative histochemical determination of succinic dehydrogenase activity in skeletal muscle fibres. Histochem J 20: 230–243

Blanco CE, and Sieck GC (1992) Quantitative determination of calcium-activated myosin adenosine triphosphatase activity in rat skeletal muscle fibres. Histochem J 24: 431–444

Bodine SC and Pierotti DJ (1996) Myosin heavy chain mRNA and protein expression in single fibers of the rat soleus following reinnervation. Neurosci Lett 215: 13–16

Bottinelli R and Reggiani C (2000) Human skeletal muscle fibres: molecular and functional diversity. Prog Biophys Mol Biol 73 (2–4): 195–262

Borgers M, Firth JA, Stoward PJ, Verheyen A (1991) Phosphatases. In: Stoward PJ, Pearse AGE (eds) Histochemistry. Theoretical and applied. Churchill Livingstone, Edinburgh, London, Melbourne, New Yorl, Tokyo, 4thedn, pp 187– 218

Buchwalow IB, Schulze W, Kostic MM, Wallukat G, Morwinski R (1996) Involvement of adenylyl cyclase in nitric oxide synthase expression by neonatal rat cardiomyocytes in vitro. Circulation 94: 3451

Buchwalow IB, Kostic MM, Schulze W, Wallukat G, Krause E-G, and Haller H (1997) Subcellular localization of inducible nitric oxide synthase in cardiomyocytes and control of its expression and activity. Circulation 96: 938

Buchwalow IB, Schulze W, Kostic MM, Wallukat G, Morwinski R (1997a) Intracellular localization of inducible nitric oxide synthase in neonatal rat cardiomyocytes in culture. Acta histochem. 99: 231–240

Buchwalow IB, Schulze W, Karczewski P, Kostic MM, Wallukat G et al. (2001) Inducible nitric oxide synthase in the myocard. Mol Cell Biochem 217: 73–82

Burke RE (1983) Motor units– anatomy, physiology and functional organization. In: Brooks VB (ed) Handbook of Physiology. The Nervous System. Vol 2, Bethesda, pp 345–422

Burnham R, Martin T, Stein R, Bell G, Maclean I, and Steadward R (1997) Skeletal muscle fibre type transformation following spinal cord injury. Spinal cord 35 (2): 86–91

Butler J, Cosmos E, Brierley J (1982) Differentiation of muscle fiber types in neurogenetic brachial muscles of the chick embryo. J Exp Zool 224: 65

Bylund A-C, Bjurö T, Cederblad G, Holm J, Lundholm K, Sjöström M, and Änquist KA (1977) Physical training in man. Skeletal muscle metabolism in relation to muscle morphology and running ability. Eur J Appl Physiol 36: 151–169

Close R (1964) Dynamic properties of fast and slow skeletal muscles of the rat during development. J Physiol (Lond) 173: 74–95

Condon K, Silberstein L, Blau HM, Thompson WJ (1990) Differentiation of fiber types in aneural musculature of the prenatal rat hind limb. Dev Biol 138: 275

Costill DL, Daniels J, Evans W, Fink W, Krahenbuhl G, and Saltin B (1976a) Skeletal muscle enzymes and fibre composition in male and female track athletes. J Appl Physiol 40: 149–154

Cotter MA, Cameron NE, Robertson S, and Ewing I (1993) Polyol pathway-related skeletal muscle contractile and morphological abnormalities in diabetic rats. Exp Physiol 78: 139–155

Craig R (1994) The structure of contractile filaments. In: Engel AG, Franzini-Armstrong C (eds) : Myology. Mc-Graw-Hill Inc, New York, St Louis, San Francisco, Auckland, Bogota, Caracas, Lisbon, London, Madrid, Mexico City, Milan, Montreal, New Dehli, Paris, San Juan, Singapore, Sydney, Tokyo, Toronto, 2nd edn, pp 134–175

Czesla M, Mehlhorn G, Fritzsche D, and Asmussen G (1997) Cardiomyoplasty– improvement of muscle fibre type transformation by an anabolic steroid (metenolone). J Mol Cell Cardiol 29: 2989–2996

Daugaard JR, Laustsen JL, Hansen BS, Richter EA (1998) Growth hormone induces muscle fibre type transformation in growth hormone-deficient rats. Acta physiol Scand 164 (2): 119–126

Denardi C, Ausoni S, Moretti P, Gorza L, Yelleca M, Buckingham M, and Schiaffino S (1993) Type 2X-myosin heavy chain is coded by a muscle fiber type-specific and developmentally regulated gene. J Cell Biol 123: 823–835

Drieu K (1988) Preparation and definition of Ginkgo biloba extract. In: Fünfgeld EW (ed) Rökan Ginkgo biloba. Recent results in pharmacology and clinic. Springer, Berlin, Heidelberg, New York, London, Paris, Tokyo, pp 31–36

Dyck PJ (1994) Diseases of peripheral nerves. In: Engel AG and Franzini-Armstrong C (eds) Myology, McGraw-Hill Inc, New York, St Louis, San Francisco, Auckland, Bogota, Caracas, Lisbon, London, Madrid, Mexico City, Milan, Montreal, New Dehli, Paris, San Juan, Singapore, Sydney, Tokyo, Toronto, 2ndedn., pp 1870–1904

Ebert D und Asmussen G (1986) Die Rolle der motorischen Einheiten bei der Steuerung von muskulären Aktionen. Z Physiother 38: 287–304

Edgerton VR, Roy RR, Bodine-Fowler SC, Pierotti DJ, Unguez GA, Martin TP (1990) Motoneurons-Muscle fibre connectivity and interdependence. In: Pette D (ed) The Dynamic State of Muscle Fibers. Walter de Gruyter, Berlin, New York, pp 217–232

Edström L, and Kugelberg E (1968) Histochemical composition, distribution of fibres and fatigability of single motor units. J Neurol Neurosurg Psychiat 31: 424–433

Engel AG, Franzini-Armstrong C /eds/ (1994): Myology. Mc-Graw-Hill Inc, New York, St Louis, San Francisco, Auckland, Bogota, Caracas, Lisbon, London, Madrid, Mexico City, Milan, Montreal, New Dehli, Paris, San Juan, Singapore, Sydney, Tokyo, Toronto, 2nd edn.

Eriksson KF, Saltin B, and Lindgarde F (1994) Increased skeletal muscle capillary density precedes diabetes development in men with impaired glucose tolerance: A 15-yeat follow-up. Diabetes 43: 805–808

Förstermann U, and Kleinert H (1995) Nitric oxide synthase: expression and expressional control of the three isoforms. Naunyn-Schmiedeberg`s Arch Pharmacol 352: 351–364

Frandsen U, Lopez-Figueroa M, Hellsten Y (1996) Localization of nitric oxide synthase in human skeletal muscle. Biochem Biophys Res Commun 227 (1): 88–93

Franzini-Armstrong C and Fischman DA (1994) Morphogenesis of skeletal muscle fibers. In: Engel AG, Franzini-Armstrong C (eds) Myology. Mc-Graw-Hill Inc, New York, St Louis, San Francisco, Auckland, Bogota, Caracas, Lisbon, London, Madrid, Mexico City, Milan, Montreal, New Dehli, Paris, San Juan, Singapore, Sydney, Tokyo, Toronto, 2nd edn., pp 74–96

Fredette BJ and Landmesser L (1991) A reevaluation of the role of innervation in primary and secondary myogenesis in developing chick muscle. Dev Biol 143: 19

Gath I, Closs EI, Godtel-Armbrust U, Schmitt S, Nakane M, Wessler I, Förstermann U (1996) Inducible NO synthase II and neuronal NO synthase I are constitutively expressed in different structures of guinea pig skeletal muscle: implications for contractile function. FASEB J 10 (14): 1614–1620

Gath I, Ebert J, Godtel-Armbrust U, Ross R, Reske-Kunz AB, Förstermann U (1999) NO synthase II in mouse skeletal muscle is associated with caveolin 3. Biochem J 340: 723–728

Gauthier GF (1990) Differential distribution of myosin isoforms among the myofibrils of individual developing muscle fibers. J Cell Biol 110: 693–701

Gerber G and Siems W (1987) Äthiopathogenetische Bedeutung von aktivierten Spezies des Sauerstoffs (O$_2$-Radikale) bei Hypoxie und Ischämie. Z Klein Med 42: 1025–1029

Gosselin LE, Zhan W-Z, and Sieck GC (1996) Hypothyroid-mediated changes in adult rat diaphragm muscle contractile properties and MHC isoform expression. J Appl Physiol 80: 1934–1939

Griggs RC and Markesbery WR (1994) Distal myopathies. In: Engel AG, Franzini-Armstrong C (eds) Myology. Vol 2 Mc-Graw-Hill Inc, New York, St Louis, San Francisco, Auckland, Bogota, Caracas, Lisbon, London, Madrid, Mexico City, 2nd edn, p1246

Grossi J (1982) Contractile and electrical characteristics of extensor digitorum longus muscle from alloxan-diabetic rats. An in vitro study. Diabetes 31: 194–202

Grozdanovic Z and Gossrau G (1997) Nitric oxide synthase (NOS) I during postnatal development in rat and mouse skeletal muscle. Acta histochem 99: 311–324

Grozdanowic Z, Christova T, and Gossrau R (1997) Differences in the localization of the postsynaptic nitric oxide synthase I and acetylcholinesterase suggest a heterogeneity of neuromuscular junctions in rat and skeletal muscles. Acta histochem 99: 47–53

Guillon JM, Rochette L, and Baranes J (1988) Effects on Ginkgo biloba extract on two models of experimental myocardial ischemia. In: Fünfgeld EW (ed) Rökan Ginkgo biloba. Recent results in pharmacology and clinic. Springer, Berlin, Heidelberg, New York, London, Paris, Tokyo, pp 153–161

Guth L (1973) Fact and artifact in the histochemical procedure for myofibrillar ATPase. Exp Neurol 41: 440–448

Guth L and Samaha F (1969) Qualitative differences between actomyosin ATPase of slow and fast mammalian mucsle. Exp neurol 25: 138–144

Guth L, and Yellin H (1971) The dynamic nature of the so-called " fibre types" of mammalian skeletal muscle. Exp Neurol 31: 277–285

Guth L and Samaha F (1972) Erroneous interpretations which may result from application of the " myofibrillar ATPase" histochemical procedure to developing muscle. Exp Neurol 34: 465–475

103

Halkajer-Kristinsen K and Ingemann-Hansen T (1981) Variations in single fibre areas and fibre composition in needle biopsies from the human quadriceps muscle. Scand J Clin Lab Invest 41: 391–395

Harker DW (1972) The structure and innervation of sheep superior rectus and levator palpebrae extraocular muscles I. Extrafusal muscle fibers. Invest Ophthalmol 11: 956–969

Henriksson-Larsen, Friden J, and Wretling ML (1985) Distribution of fibre sizes in human skeletal muscle. An enzyme histochemical study in m. tibialis anterios. Acta Physiol Scand 123: 171–177

Hochachka PW (1992) Muscle enzymatic composition and metabolic regulation in high altitude adapted natives. Int J Sports Med 13 (Suppl 1): 89–91

Hoppeler H (1990) The range of mitochondrial adaption in muscle fibers. In: Pette D (ed) The Dynamic State of Muscle Fibers. Walter de Gruyter, Berlin, New York, pp 567–586

Hoppeler H and Desplanches D (1992) Muscle structural modifications in hypoxia. Int J Sports Med 13 (Suppl 1): 166–168

Hori A, Ishihara A, Kobayashi S, and Ibata Y (1998) Immunohistochemical classification of skeletal muscle fibers. Acta Histochem Cytochem 31 (5) 375–384, p 381

Howald H, Pette D, Simoneau JA, Hoppeler H, and Cerretelli P (1990) Effect on chronic hypoxia on muscle enzyme activities. Int J Sports Med 11 (Suppl 1): 10–14

Huxley AF and Niedergerke R (1954) Structural changes in muscle during contraction. Interference microscopy of living muscle fibres. Nature 173: 971–973

Huxley HE and Hanson J (1954) Changes in the cross-striations of muscle during contraction and stretch and their structural interpretation. Nature 173: 973–976

Ishihara A, Roy RR, and Edgerton VR (1995) Succinate dehydrogenase activity and soma size of motoneurons innervating different portions of the rat tibialis anterior. Neuroscience 68 (3): 813–822

Jansson E, Sjödin B, and Tesch P (1978) Changes in muscle fibre type distribution in man after physical training. A sign of fibre transformation? Acta Physiol Scand 104: 235–237

Jostarndt K, Puntschart A, and Hoppeler H (1996) Fiber Type-specific expression of essential (alkali) myosin light chains in human skeletal musle. Histochem J Cytochem 44: 1141–1152

Josza L, Vandor E, Demel S, Rapcsak M, Reffy A, Szoor A, and Hideg J (1985) Histochemical and biochemical alterations in skeletal muscles of rats during combined chronic hypoxia and hypokinesia. Gegenbaurs Morphol Jahrb 131: 43–54

Kainulainen H, Komulainen J, Joost HG, and Vihko V (1994) Dissociation of the effects of training on oxidative metabolism. Glucose utilisation and GLUT4 levels in skeletal muscle of streptozotocin-diabetic rats. Pflügers Arch 427: 444–449

Kapur S, Bedard S, Marcotte B, Cote CH, Marette A (1997) Expression of nitric oxide synthase in skeletal muscle: a novel role for nitric oxide as a modulator of insulin action. Diabetes 46 (11): 1691–700

Kato T (1938) Über histologische Untersuchungen der Augenmuskeln von Menschen und Säugetieren. Okajimas Folia anat Jap 16: 131–145

Kelly AM and Rubinstein NA (1994) The diversity of muscle fiber types and its origin during development. In: Engel AG and Franzini-Armstrong C (eds) Myology. McGraw-Hill,Inc., New York, St Louis, San Francisco, Auckland, Bogota, Caracas, Lisbon, London, Madrid, Mexico City, Milan, Montreal, New Dehli, Paris, San Juan, Singapore, Sydney, Tokyo, Toronto, 2ndedn, pp119–133

Khuchua ZA, Ventura-Clapier R, Kuznetsov AV et al. (1989) Alterations in the creatine kinase system in the myocardium of cardiomyopathic hamsters. Biochem Biophys Res Commun 165: 748–157

Kirshenbaum LA and Singal PK (1992) Changes in antioxidant enzymes in isolated cardiac myocytes subjected to hypoxia-reoxygenation. Lab Invest 67: 796–803

Klueber KM and Feczko JD (1994) Ultrastructural, histochemical, and morphometric analysis of skeletal muscle in a murine model of type I diabetes. Anat Rec 239: 18–34

Knudsen KA and Horwitz AF (1994) The plasma membrane of the muscle fibre: adhesion molecules. In: Engel AG, Franzini-Armstrong C (eds) Myology. Mc-Graw-Hill Inc, New York, St Louis, San Francisco, Auckland, Bogota, Caracas, Lisbon, London, Madrid, Mexico City, Milan, Montreal, New Dehli, Paris, San Juan, Singapore, Sydney, Tokyo, Toronto, 2nd edn., pp 223–241

Kobzik L, Reid MB, Bredt DS, and Stamler JS (1994) Nitric oxide in skeletal muscle. Nature 372, 546–548

Kobzik L, Stringer B, Balligand J-L, Reid MB, and Stamler JS (1995) Endothelial type nitric oxide synthase in skeletal muscle fibers: mitochondrial relationships. Biochem Biophys Res Commun 211 (2) 375–381

Kjeldsen K, Bjerregaard P, Richter EA, et al. (1988) Na+, K+ ATPase concentration in rodent and human heart and skeletal muscle: apparent relation to muscle performance. Cardiovasc Res 22: 95–100

Krug H (1980) Histo- und Zytophotometrie. Gustav Fischer, Jena, pp12–22

Kugelberg E and Lindegren B (1979) Transmission and contraction fatigue of rat motor units in relation to succinate dehydrogenase activity of motor unit fibres. J Physiol 288: 285–300

Kuo TH, Tsang W, Wang KK et al. (1992) Simultaneous reduction of the sarcolemmal and SR calcium ATPase activities and gene expression in cardiomyopathic hamster. Biochem Biophys Acta 1138: 343–349

Laemmli UK (1970) Cleavage of structural proteins during the assembly of the head of bacteriophage T4. Nature 227: 680–685

Larsson L (1992) Is the motor unit uniform? Acta Physiol Scand 144: 143–154

Larsson L, Edström L, Lindegren B, Gorza L, and Schiaffino S (1991) Myosin heavy chain composition and enzyme- histochemical and physiological properties of a novel fast-twitch motor unit type. Am J Physiol 261: C93–C101

Leigh RJ, Zee DS (1991) The Neurology of Eye Movements. Davis, Philadelphia

Li ZB, Rossmanith GH, Hoh JFY (2000) Cross-bridge kinetics of rabbit single extraocular and limb muscle fibres. Invest Ophthalmol 41: 3770–3774

Lojda Z, Gossrau R, and Schiebler TH (1976) Enzymhistochemische Methoden. Springer, Berlin, Heidelberg, New York, p 249

Lucas CA, Rughani A, Hoh JFY (1995) Expression of extraocular myosin heavy chain in rabbit laryngeal muscle. J Muscle Res Cell Motil 16: 368–378

Malhorta A, Karell M, Scheuer J (1985) Multiple cardiac contractile protein abnormalities in myopathic Syrian hamsters (BIO 53:58) J Mol Cell Cardiol 17: 95–107

Martin TP, Bodine-Fowler S, Roy PR, Eldred E, and Edgerton VR (1988a) Metabolic and fibre size properties of rat tibialis anterior motor units. Am J Physiol 255: C43–C50

Martin TP, Bodine-Fowler S, and Edgerton VR (1988b) Coordination of electromechanical and metabolic properties of cat soleus motor units. Am J Physiol 255: C684–C693

Mayr R (1971) Structure and distribution of fibre types in the external eye muscles of the rat. Tissue Cell 3: 433–462

Mayr R (1973) Morphometrie von Ratten-Augenmuskelfasern. Verh Anat Ges 67: 353–358

Mayr R, Gottschall J, Gruber H, Neuhuber W (1975) Internal structure of cat extraocular muscle. Anat Embryol 148: 25–34

Mc Cabe ERB (1994) Microcompartmentation of energy metabolism at the outer mitochondrial membrane: Role in diabetes mellitus and other diseases. J Bioenerg and Biomembr 26: 317–325

Meißner JD, Kubis H-P, Scheibe RJ, and Gros G (2000) Reversible Ca 2+-induced fast-to-slow transition in primary skeletal muscle culture cells at the mRNA level. J Physiol 523: 19–28

Mohr W, Lossnitzer K (1974) Morphologische Untersuchungen an Hamstern des Stammes BIO 8262 mit erblicher Myopathie und Kardiomyopathie. Beitr Path 153: 178–193

Morgan DL, Proske U (1984) Vertebrate slow muscle: its structure, pattern of innervation, and mechanical properties. Physiol Rev 64: 103–169

Moss FP and Leblond CP (1971) Satellite cells as the source of nuclei in muscles of growing rats. Anat Rec 170: 421–436

Nakano H, Masuda K, Sasaki S, and Katsuta S (1997) Oxidative enzyme activity and soma size in motoneurons innervating the rat slow-twitch and fast-twitch muscles after chronic activity. Brain Res Bull 43 (2): 149–54

Narusawa M, Fitzsimons RB, Izumo S, Nadal-Ginard, B, Rubinstein NA, and Kelly AM (1987) Slow myosin in developing rat skeletal muscle. J Cell Biol 104: 447–259

Nelson JS, Goldberg SJ and McClung JR (1986) Motoneuron electrophysiological and muscle contractile properties of superior oblique motor units in cat. J Neurophysiol 55: 715–726

Nemeth PM, Hofer H, and Pette D (1979) Metabolic heterogeneity of muscle fibers classified by myosin ATPase. Histochemistry 63: 191–199

Norgaard A, Baandrup U, Larsen JS et al. (1987) Heart Na, K-ATPase activity in cardiomyopathic hamsters as estimated from K-dependent 3-0-MFPase activity in crude homogenates. J Mol Cell Card 19: 589–594

Ovalle WK and Smith RS (1972) Histochemical identification of three types of intrafusal muscle fibres in the cat and monkey based on the myosin ATPase reaction. Can J Physiol Pharmacol 50, 195–201

Padykula HA, and Herman E (1955) Factors effecting the activity of adenosine triphosphatase and other phosphatases as measured by histochemical techniques. Histochem J cytochem 3: 161–169

Pai VD, Macha N, and Curi PR (1982) Cytophotometric study of succinate dehydrogenase activity on individual skeletal muscle fibres of the rat. Gegenbaurs morph Jahrb 128: 201–207

Park CS, Park R, Krishna G (1996) Constitutive expression and structural diversity of inducible isoform of nitric oxide synthase in human tissues. Life Sci 59 (3): 219–25

Pastoris O, Vercaesi L, and Dosena M (1991) Effects of hypoxia and pharmacological treatment on enzyme activities in skeletal muscle of rats of different ages. Exp Geront 26: 77–87

Paulus SF and Grossie J (1983) Skeletal muscle in alloxan diabetes. A comparison of isometric contractions in fast and slow muscle. Diabetes 32: 1035–1039

Pedrosa-Domelöff F, Erikson PO, Butler-Browne GS, Thornell LE (1992) Expression of alpha-cardiac myosin heavy chain in mammalian skeletal muscle. Experientia 48: 491–494

Peter JB, Barnard RJ, Edgerton VR, Gillespie CA, and Stempel KE (1972) Metabolic profiles of three fiber types of skeletal muscle in guinea pigs and rabbits. Biochemistry 11: 2627–2633

Pette D (1966) Mitochondrial enzyme activities. In: Tager JM, Papa S, Quagliariello E, Slater EC (eds) Regulation of metabolic processes in mitochondria. Vol 2, Elsevier, Amsterdam London New York, pp 28–50

Pette D (1984) Activity-induced fast to slow transitions in mammalian muscle. Med Sci Sports Exerc 16: 517–528

Pette D (1990a) The Dynamic State of Muscle Fibers. Walter de Gruyter, Berlin, New York

Pette D (1990b) Dynamics of stimulation-induced fast-to-slow transitions in protein isoforms of the thick and thin filament. In: Pette D (ed) The Dynamic State of muscle Fibers. Walter de Gruyter, Berlin, New York, pp 415–428

Pette D (1998) Training effects on the contractile apparatus. Acta Physiol Scand 162: 367–376

Pette D, and Bücher T (1963) Proportionskonstante Gruppen in Beziehung zur Differenzierung der Enzymaktivitätsmuster von Skelett- Muskeln des Kaninchens. Hoppe-Seyler´s Z Physiol Chem 331: 180–195

Pette D, and Dölken G (1975) Some aspects of regulation of enzyme levels in muscle energy-supplying metabolism. In: Weber G (ed) Advances in enzyme regulation. Vol 13, Pergamon Press- Oxford and New York, pp 355–377

Pette D and Staron RS (1997) Mammalian skeletal muscle fibre type transitions. Int Rev Cytol 170: 143–223

Pette D, and Vrbova G (1985) Neuronal control of phenotypic expression in mammalian muscle fibres. Muscle and Nerve 8: 676–689

Peuker H and Pette D (1997) Quantitative analyses of myosin heavy-chain mRNA and protein isoforms in single fibers reveal a pronounced fiber heterogeneity in normal rabbit muscles. Eur J Biochem 247: 30–36

Phillips WD and Bennet MR (1984) Differentiation of fiber types in wing muscles during embryonic development. Effect of neural tube removal. Dev Biol 106: 457–467

Pieper KS (1984b) Muskulatur. In: Grundlagen der Sportmedizin: Nervensystem, Muskulatur, Stoffwechsel. Lehrheft 1, Deutsche Hochschule für Körperkultur Leipzig, 1. Auflage, pp 48–85

Pieper KS, Seidler E, Bähr B, Förster E, Paul I, and Gärtner CH (1984a) Zur Wechselbeziehung zwischen dem histochemischen Nachweis der Glyzerin-1-Phosphatoxidase und dem LDH-Isoenzymmuster im Skelettmuskel. Acta histochem Suppl. XXX, 257–263

Pierobon-Bormioli S, Sartore S, Vitadello M, Schiaffino S (1980) "Slow" myosins in vertebrate skeletal muscle. J Cell Biol 85: 672–681

Pincemail J and Deby C (1988) The antiradical properties of Ginkgo biloba extract. In: Fünfgeld EW (ed) Rökan Ginkgo biloba. Recent results in pharmacology and clinic. Springer, Berlin, Heidelberg, New York, London, Paris, Tokyo, pp 71–82

Polgar J, Johnson MA, Weightman D, and Appleton D (1973) Data on fibre size in thirty-six human muscles. An autopsy study. J Neurol Sci 19: 307–318

Porter JD, Baker RS, Ragusa RJ, Brueckner JK (1995) Extraocular muscles: basic and clinical aspects of structure and function. Surv Ophthalmol 39: 451–484

Prior BM, Ploutz-Snyder LL, Cooper TG, Meyer RA (2001) Fibre type and metabolic dependence of T2 increases in stimulated rat muscles. J Appl Physiol 90: 615–623

Punkt K and Erzen I (2000) Changes of enzyme activities in the myocardium and skeletal muscle fibres of cardiomyopathic hamsters. A cytophotometrical study. Exp Toxic Pathol 52: 103–110

Punkt K, Punkt J, Krug H, and Schippel G (1984) Comparison of histophotometrical and biochemical myosin-ATPase estimations. Histochem J 16: 385–387

Punkt K, Krug H, and Taubert G (1985) A new system to consider varying section thickness in histophotometry. Acta histochem 76: 209–212

Punkt K, Krug H, Punkt J, Erzen I, and Meznaric M (1986) The correlation between histohotometrical and biochemical myosin-ATPase measurements in the myocardium and striated muscle of the rat. Acta histochem 78: 105–109

Punkt K, Krug H, and Böhme R (1987) Histophotometrical measurements concerning the distribution of myofibrillar ATPase activity within the tissue block after aldehyde fixation. Acta histochem 82: 109–113

Punkt K, Krug H, Huse J, and Punkt J (1993) Age-dependent changes of enzyme activities in the different fibre types of of rat extensor digitorum longus and gastrocnemius muscles. Acta histochem 95: 97–110

Punkt K, Welt K, and Schaffranietz L (1995) Changes of enzyme activities in the rat myocardium caused by experimental hypoxia with and without Ginkgo biloba extract EGb 761 pretreatment. Acta histochem 97: 67–79

Punkt K, Unger A, Welt K, Hilbig H, and Schaffranietz L (1996) Hypoxia-dependent changes of enzyme activities in different fibre types of rat soleus and extensor digitorum longus muscles. A cytophotometrical study. Acta histochem 98: 255–269

Punkt K, Adams V, Linke A, and Welt K (1997) The correlation of cytophotometrically and biochemically measured enzyme activities: Changes in the myocardium of diabetic and hypoxic diabetic rats, with and without Ginkgo biloba extract treatment. Acta histochem 99: 291–299

Punkt K, Mehlhorn H, and Hilbig H (1998) Region- and age-dependent variations of muscle fibre properties. Acta histochem 100: 37–58

Punkt K, Psinia I, Welt K, Barth W, and Asmussen G (1999) Effects on skeletal muscle fibres of diabetes and Ginkgo biloba extract treatment. Acta histochem 101: 53–69

Punkt K, Zaitsev S, Park JK, Wellner M, and Buchwalow IB (2001) Nitric oxide synthase-isoforms I, III and protein kinase-C θ in skeletal muscle fibres of normal and streptozotocin-induced diabetic rats with and without Ginkgo biloba extract-treatment. Histochem J, 33: 213–219

Ranvier L (1873) Proprietes et structures differentes des muscles rouges et des muscles blancs, chez les lapins et chez les raies. C R Acad Sci (Paris) 77: 1030–1034

Reichmann H, and Pette D (1982) A comparative microphotometric study of succinate dehydrogenase activity levels in type I, IIA and IIB fibres of mammalian and human muscles. Histochemistry 74: 27–41

Reichmann H, and Pette D (1984) Glycerolphosphate oxidase and succinate dehydrogenase activities in IIA and IIB fibres of mouse and rabbit tibialis anterior muscles. Histochemistry 80: 429–433

Reid MB (1998) Role of nitric oxide in skeletal muscle: synthesis, distribution and functional importance. Acta Physiol Scand 162 (3): 401–409

Rösen P, Windeck P, Zimmer HG, Frenzel H, Bürring KF, and Reinauer H (1986) Myocardial performance and metabolism in non-ketotic, diabetic rat hearts: myocardial function and metabolism in vivo and the isolated perfused heart under the influence of insulin and octanoate. Basic Res Cardiol 81: 620–635

Rösen P, Pogatsa G, Tschöpe D, Addicks K, and Reinauer H (1992) Diabetische Kardiopathie. Klin Wochenschr 69 (Suppl XXIX): 3–15

Rotenberg SA (1999) Roles of protein kinase C in the synthesis and cellular action of nitric oxide. In: Laskin JD and Laskin DL (eds) Cellular and molecular biology of nitric oxide. Marcel Dekker, Inc, New York, Basel, pp 171–198

Rowlerson AM (1987) Fibre types in extraocular muscles. In: Transactions of 16th Meeting of the European Strabismological Association, ed Kaufmann H, Giessen: ESA, pp 19–26.

Roy RR, Ishihara A, Kim JA, Lee M, Fox K, and Edgerton VR (1999) Metabolic and morphological stability of motoneurons in response to chronically elevated neuromuscular activity. Neuroscience 92(1): 361–366

Rubinstein NA, Hoh JFY (2000) The distribution of myosin heavy chain isoforms among rat extraocular muscle fibre types. Invest Ophthalmol 41: 3391–3398

Rushbrook JI, Weiss C, Ko K (1994) Identification of alpha-cardiac myosin heavy chain mRNA and protein in extraocular muscle of the adult rabbit. J Muscle Res Cell Motil 15:
505–515.

Sartore S, Mascarello F, Rowlerson A, Gorza L, Ausoni S, Vianello M, Schiaffino S (1987) Fibre types of extraocular muscles: a new myosin isoform in the fast fibres. J Muscle Res Cell Motil 8: 161–172

Schantz P, Randall-FOX E, Hutchison W, Tyden A, and Astrand P-O (1981) Muscle fibre type distribution, muscle cross-sectional area and maximal voluntary strength in humans. Acta Physiol Scand 117: 219–226

Schantz P, Sjoberg B, Widebeck AM, Ekblom B (1997) Skeletal muscle of trained and untrained paraplegics and tetraplegics. Acta Physiol Scand 161: 31–39

Schiaffino S and Reggiani C (1996) Molecular diversity of myofibrillar proteins: gene regulation and functional significance. Physiol Rev 76: 371–423

Schiaffino S, Gorza L, Ausoni S, Bottinelli R, Reggiani C, Larsson L, Gundersen K, and Lömo T (1990) Muscle fibre Types Expressing Different Myosin heavy Chain Isoforms. Their Functional properties and Adaptive Capacity. In: Pette D (ed) The Dynamic State of Muscle Fibers. Walter de Gruyter, Berlin, New York, pp 329–342

Schiebler TH, Schmidt W, Zilles K (eds) Anatomie. Springer, Berlin Heidelberg New York Barcelona, Hong Kong London Milan Paris Singapore Tokyo, p 74

Schmalbruch H (1985a) Skeletal muscle. Springer, Berlin Heidelberg New York Tokyo, pp 160–161

Schmalbruch H (1985b) Skeletal muscle. Springer, Berlin Heidelberg New York Tokyo, pp 181–183

Schmalbruch H (1985c) Skeletal muscle. Springer, Berlin Heidelberg New York Tokyo, pp 58–67

Seburn K, Coicou C, and Gardiner P (1994) Effects of altered muscle activation on oxidative enzyme activity in rat alpha-motoneurons. J Appl Physiol 77(5): 2269–2274

Sieck GC, Sacks RD, and Blanco CE (1987) Absence of regional differences in the size and oxidative capacity of diaphragm muscle fibers. J Appl Phys 63: 1076–1082

Simoneau JA (1990) Species-specific ranges of metabolic adaptions in skeletal muscle. In: Pette D (ed) The Dynamic State of Muscle Fibers. Walter de Gruyter, Berlin, New York, pp 587–600

Simoneau JA and Kelly DE (1998) Altered glycolytic and oxidative capacities of skeletal muscle contribute to insulin resistance in NIDDM. J Appl Physiol 83: 166–171

Silver PJ, Monteforte PB (1988) Differential effects of pharmacological modulators of cardiac myofibrillar ATPase activity in normal and myopathic (BIO 14.6) hamsters. Eur J Pharmacol 147: 335–342

Soukup T, Jirmanova I (2000) Regulation of myosin expression in developing and regenerating extrafusal and intrafusal muscle fibres with special emphasis on the role of thyroid hormones. Physiol. Res. 41: 617–683

Soukup T, Pedrosa-Domellöff F, Thornell L-E (1995) Expression of myosin heavy chain isoforms and myogenesis of intrafusal fibres in rat muscle spindles. Micr Res Tech 30: 390–407

Spencer RF, Porter JD (1988) Structural organization of the extraocular muscles. In: Neuroanatomy of the oculomotor system. ed Büttner-Ennever, Elsevier Science Publishers, Amsterdam, pp 33–79

Staron RS and Pette D (1987) Nonuniform myosin expression along single fibres of chronically stimulated and contralateral rabbit tibialis anterior muscles. Pflügers Arch 409: 67–73

Stoward PJ (1980) Criteria for the validation of quantitative enzyme histochemical techniques. In: Evered D and O´Connor M (eds) Trends in Enzyme Histochemistry and Cytochemistry. Excerpta Medica, Amsterdam, pp 11–31

Stoward PJ, Meijer AEFH, Seidler E, and Wohlrab F (1991) Dehydrogenases. In: Stoward PJ, Pearse AGE (eds) Histochemistry. Theoretical and applied. Churchill Livingstone, Edinburgh, London, Melbourne, New York, Tokyo, 4th ed, pp 27–71

Taguchi S, Hata Y, and Itoh K (1985) Enzymatic responses and adaptions to swimming training and hypobaric hypoxia in postnatal rats. Jpn J Physiol 35: 1023–1032

Takahashi H, Kikuchi K, and Nakayama H (1993) Effect of chronic hypoxia on oxidative enzyme activity in rat skeletal muscle. Ann Physiol Anthropol 12: 363–369

Taylor AW, BachmanL (1999) The effects of endurance training on muscle fibre types and enzyme activities. Can J Appl Physiol 24: 41–53

Termin A, Staron RS, and Pette D (1990) Myosin Heavy Chain Isoforms in Single Fibres. In: Pette D (ed) The Dynamic State of Muscle Fibers. Walter de Gruyter, Berlin, New York, pp 463–472

Tittel K (1981) Beschreibende und funktionelle Anatomie des Menschen. 9.Aufl., Gustav Fischer, Jena

Torre M (1953) Nombre et dimension des unites motorices dans les muscles extrinseques de l'oeil et, en general, dans les muscles squelettiques relies a des organes de sens. Schweiz Arch Neurol 72: 362–367

Towbin T, Stahelin T, and Gordon J (1979) Electrophoretic transfer of proteins from polyacrylamide gels to nitrocellulase sheets: Procedure and applications. Proc Natl Acad Sci USA 76: 4350–4354

Van der Laarse WJ, Diegenbach PC, and Maslam S (1984) Quantitative histochemistry of three mouse hind-limb muscles: the relationship between calcium-stimulated myofibrillar ATPase and succinate dehydrogenase activities. Histochem J 16: 529–541

Van der Laarse WJ, Diegenbach PC, and Hemminga MA (1986) Calcium-stimulated myofibrillar ATPae activity correlates with shortening velocity of muscle fibres in Xenopis laevis. Histochem J 18: 487–496

Van der Laarse WJ, Diegenbach PC, and Elzinga G (1989) Maximum rate of oxygen consumption and quantitative histochemistry of succinate dehydrogenase in single muscle fibres of Xenopus laevis. J Muscle Res Cell Motil 10: 221–228

Van Noorden CJF and Butcher RG (1991). Quantitative enzyme histochemistry. In: Stoward

PJ, Pearse AGE (eds) Histochemistry. Theoretical and applied. Churchill Livingstone, Edinburgh, London, Melbourne, New Yorl, Tokyo, 4th ed, pp 355–432

Van Noorden CJF and Frederiks WM (1992) Enzyme Histochemistry: A Laboratory Manual of Current Methods. Oxford University Press, Royal Microscopy Society, pp 9–19

Van Noorden CJF and Jonges GN (1995) Analysis of enzyme reactions in situ. Histochem J 27: 101–118

Veitch K, Hombroeckx A, and Hue L (1991) Ischemia/anoxia-induced increase in cardiac mitochondrial respiration: changes in Complex I. Biochem Soc Trans 19 (3): 261–268

Vetter C, Reichmann H, and Pette D (1984) Microphotometric determination of enzyme activities in type-grouped fibres of reinnervated rat muscle. Histochemistry 80: 347–351

Vrbova G, Lowrie MB, and Connold AL (1990) Motor activity dependent muscle fibre transformation of the rat soleus. In: Pette D (ed) The Dynamic State of muscle Fibers. Walter de Gruyter, Berlin, New York, pp 205–216

Walro JM, Kucera J (1999) Why adult mammalian intrafusal and extrafusal fibres contain different myosin heavy-chain isoforms. Trends Neurosci 22: 180–184

Walton J, Karpati G, and Hilton-Jones D /eds/(1994) Disorders of voluntary muscle. Churchill Livingstone, Edinburgh, London, Madrid, Melbourne, New York, Tokyo, 6th edn

Wieczorek DF, Periasamy M, Butler-Browne GS, Whalen RG, Nadal-Ginard, B. (1985): Co-expression of multiple myosin heavy chain genes, in addition to a tissue specific one in extraocular musculature. J. Cell Biol. 101, 618–629

Windisch A, Gundersen K, Szaboles MJ, Gruber H, Lomo T (1998) Fast to slow transformation of denerved and electrically stimulated rat muscle. J Physiol 510 (2): 623–32

Subject Index

Production: Druckhaus Beltz, Hemsbach